Hunter Wine

Julie McIntyre is a Research Fellow in History at the University of Newcastle and Australia's foremost historian of wine in society and culture. Her publications include *First Vintage: Wine in Colonial New South Wales*, winner of a Gourmand Publishing Prize and shortlisted for the NSW Premier's History Awards. She is an associate editor of the *Journal of Wine Research*.

John Germov is a sociologist and author of more than 20 books, including *The Social Appetite: A Sociology of Food and Nutrition*. He is Pro-Vice Chancellor of the Faculty of Education & Arts at the University of Newcastle and leads the university's Wine Studies Research Network.

Hunter Wine

A HISTORY • JULIE McINTYRE AND JOHN GERMOV

NEWSOUTH

I'm not sure yet when
you'll have my response.
But, listen: a rake at work this early.
Above, alone in the vineyard, a man
is already talking with the earth.

Rainer Maria Rilke, *The Essential Rilke*,
selected and translated by Galway Kinnell
and Hannah Liebmann (HarperCollins,
New York, 2000), p. 71.

Who can imagine a farmer, after he has grown, harvested, thrashed and bagged his wheat, being obliged not only to grind it, but also convert it into bread before he can sell it.

Alexander Kelly, *The Vine in Australia* (Sands, Kenny and Co., Melbourne, 1861), p. 7.

Up among these Pokolbin hills, where the soil is of a rich volcanic nature, there are acres and acres of vines. They cover every crest, every slope, and as a result of generous moisture each tree held up by a stout stake, and eagerly spreading over the wire supports, is lavish in its growth ... but vine-culture is an intricate business, and until the vintage it is impossible to indicate whether the prolific yield will possess that fine, delicate flavour which has made a name for the best qualities of Australian wines ... Here some of the best hocks and clarets procurable in the States are grown.'

Sydney Morning Herald, 20 November 1906

Title page: Pokolbin, c1949. Photograph by Max Dupain.
Tulloch Family Collection, courtesy of J.Y. Tulloch.

Contents

A note on style and measurements *vi*
A timeline of the Hunter winegrowing community *xi*
The Hunter region map *xiv*
Introduction: making wine history 1

Chapter 1	*Vitis vinifera* in Aboriginal Country: introducing an ancient plant to an ancient land	26
Chapter 2	A network of instigators: imagining Hunter winegrowing, 1828–35	46
Chapter 3	Experiments and expectations: willing Hunter wine into being, 1836–46	78
Chapter 4	Creating an industry through co-operation: the Hunter River Vineyard Association, 1847–76	102
Chapter 5	A common people's paradise: German immigration from 1849	136
Chapter 6	New worlds of wine, at home and abroad: first wave globalists, 1877–1900	158
Chapter 7	Amid turmoil and temperance: the forgotten generation and the stoics, 1901–54	184
Chapter 8	Civilised drinking in a wine nation: the Renaissance generation, 1955–83	230

Conclusion: the blood of the grape 268
Notes 272
Appendices 282
Bibliography 288
Acknowledgments 294
Index 298

A note on style and measurements

OUR TIMEFRAME CROSSES OVER FROM THE USE OF IMPERIAL MEASUREMENTS in Australia since colonisation, to the adoption in 1966 of a metric system. We have recorded original measurements in source material and offer this conversion table.

Measurements
1 gallon = 4.55 litres; 26.5 gallons = 1 hectolitre
1 gallon = six bottles of wine of 750 millilitres
1 acre = 0.405 hectares
1 mile = 1.6 kilometres

A *hogshead* is a hooped wooden barrel used for storing and transporting wine and other liquids. Barrels come in many sizes, but hogsheads were a common colonial mode and measure.

1 hogshead = approximately 66 gallons of wine
1 barriques = 50 gallons and 1 puncheon = 110 gallons

Grape and wine nomenclature

NAMES OF WINE GRAPE VARIETALS OR CULTIVARS, AND NAMES OF WINE styles, are integral to wine history as portals to past vineyard practices and wine tastes. The coalescence in recent history of standards in grape and wine nomenclature is a window to the distinctive workings of the Hunter and Australian wine industry. But the very fact that the recorded names for wine grape varietals and Australian wine styles have evolved over time in response to producer, market and consumer pressures means there is no stable nomenclature for either vines or wine across the time period from the 1820s to the 1980s.

For example, the same cultivar of *Vitis vinifera* might equally have been called a red Hermitage in one document record from the 19th century, a Red Hermitage in another, a Shiraz, shiraz or scyras. Although just because a grower records that a grape variety was Red Hermitage does not mean the vineyard referred to unequivocally contained Shiraz grapes, or that the Red Hermitage wine contained only Shiraz grapes. Every effort has been made not to use nomenclature without the support of evidence.

To achieve uniformity of grape and wine descriptions in this book, the names of varietals and styles are capitalised, except in direct quotations from documents where there are lower case names, and for wine names commonly uncapitalised, such as port.

Grape yields

WE HAVE NOT ENGAGED DIRECTLY WITH NOTIONS OF IDEAL WINE GRAPE yields, as concepts of quantity versus quality of wine grapes gained per acre/hectare changed during our time period. Instead we offer this perspective:

> A ton of white grapes can produce about 150 gallons of wine, a ton of black grapes 100 to 130 gallons of red wine, with the thin-skinned whites at the top of that scale, and the austere Cabernet Sauvignon at the bottom, all pip and skin, and tough skin at that ... Most growers are happy with 300 gallons [of wine] consistently annually off each acre [of grapevines, where there are approximately 900 vines per acre] ...
>
> MAX LAKE IN WS PARKES, JIM COMERFORD, MAX LAKE, *MINES, WINES AND PEOPLE: A HISTORY OF GREATER CESSNOCK* (CITY OF GREATER CESSNOCK, CESSNOCK, 1979), P. 232.

> Similarly, it is thought that the optimal planting of vines is 1500 to 2222 vines/hectare (or 600 to 899 vines per acre). A vineyard investment might expect to begin to have (neutral or) positive cash flow in the fourth year after planting, and break even from between seven and ten years (depending on grape prices). Premium grapes in cool climate yield approximately 12.5 tonnes per hectare. Vines yield at a mature level after 10 years.
>
> DIANNE DAVIDSON, *A GUIDE TO GROWING WINEGRAPES IN AUSTRALIA*, 2ND EDITION (DIANNE DAVIDSON CONSULTING SERVICES, HAHNDORF, 1992), PP. 113–15.

Pokolbin, c1949.
Photograph by Max Dupain.
Tulloch Family Collection,
courtesy of J.Y. Tulloch.

A timeline of the Hunter winegrowing community

c 65 000 YEARS AGO–1827

c 65 000 Before the present era Aboriginal Australians arrive

1787 *Vitis vinifera* for the British colony of New South Wales purchased from winegrowers at the Dutch colony of the Cape of Good Hope

1788 Colonial invasion of Aboriginal Country in and around Sydney, vines planted at various sites in Sydney as a part of settler occupation

1792 Wine made from vines at Parramatta near Sydney

1797–1804 Earliest invasion of Aboriginal lands that colonists named Hunter's River

1812 Emancipist settlers establish farms in the Paterson district of the Hunter region

1816 Botanic Gardens opens in Sydney to supply imported plants for settlers and to collect Indigenous flora to advance knowledge of biota

1821 Settler capitalists encouraged to invest in land in the Hunter Valley

1825 Granting of Hunter Valley land for the Australian Agricultural Company

1826 Formation of the first mounted police force in New South Wales, intended to subdue Aboriginal insurgents

Up to 1827 Several further importations of *V. vinifera* stocks

1827 Formation of Hunter's River Agricultural Club

[Emptying barrels of grapes into a hopper, Mount Pleasant winery, New South Wales, 1950]. Photograph by Max Dupain. National Library of Australia, PIC/8729/10 LOC Album 757.

1828–46 — THE IMAGINERS

From 1828 Early vine plantings in the Hunter region

1830 Botanic Gardens distributes vine parcels of 50 cuttings to applicants

1832 Wine made at the Australian Agricultural Company HQ at Tahlee

1833–38 Importation of vines from Europe continues

1840 Outset of economic depression

1843 Statistics begin to be collected on colonial winegrowing

1843 *Maitland Mercury* the first newspaper outside of Sydney region

1846 Earliest recorded export of wine from the Hunter, to Sydney

1847–76 — THE EXPERIMENTERS

1847–76 Life of the Hunter River Vineyard Association

1849 Immigrants begin arriving from southwestern German states to bolster winegrowing, usually sponsored by settler capitalists

1860s Powdery mildew arrives in Australia and reaches the Hunter region

1860s Conditions achieved for a colonial wine industry in New South Wales

1861 Land reforms allowing smaller size farms at affordable terms

1862 *Sale of Colonial Wine Act*

1865 Practice of espaliering grape vines in the Hunter first recorded

1877–1900 — FIRST WAVE GLOBALISTS

1877 Phylloxera detected in other parts of Australia, but not in the Hunter

1877–1900 Continuing tradition of grape plantings for wine (very small table grape industry), reputation established for light style table wines; exports to Britain, Europe and the Pacific

1882 Bordeaux Exhibition, France, a triumph (one of many exhibitions in the era at local, colonial, intercolonial and international levels)

1890 Outset of economic Depression

1893 Floods spell the end of winegrowing in areas prone to inundation

1901–30 — THE FORGOTTEN GENERATION

1901 Federation of the Australian Commonwealth of states from former British colonies of New South Wales, and others, creating a national wine industry and interstate free trade of wine

1901 Inauguration of the Pokolbin and District Vine Growers Association

1901 PDVGA instrumental in laws for wine purity

1914–18 Australian forces engaged in World War I

1916 Referendum introduces early closing for licenced premises

1919 Beginning of soldier settlement in the Hunter

1920s Downy mildew arrives in Australia and reaches the Hunter

1923 Laws on 'colonial wine' amended to 'Australian wine'

1929 Onset of the 'Great Depression'

1931–54 — THE STOICS

1931 Report on the Wine Industry of Australia focused chiefly on other wine regions in the nation, and predicts demise of Hunter wine region

1931–54 Number of vine growers in the region falls from 100 to 10 but wine production continues, and reputation for table wines strengthens

1939–45 Australian forces engaged in World War II

1946 Electricity available in some parts of the Hunter Valley

1947 Referendum on liquor licencing in New South Wales retains prohibition on the sale of alcohol after 6pm, six days a week

1955 Electricity connected to Pokolbin

1949–60 Operation of Greta Migrant Camp

1951–54 Establishment, hearing and reporting of the Royal Commission on Liquor Laws in New South Wales

1955–83 — THE RENAISSANCE GENERATION

1955 Floods devastate the Lower Hunter

1955 Referendum ends early closing for licenced premises, wine can be served after 6pm in restaurants and other places, beginning of new era of 'civilised drinking'

1959 Lowest ebb in Hunter vineyard acreage in the 20th century (approx. 500 acres)

From early 1960s vine plantings begin again and new 'introduction' table wine styles gain popularity among women as well as men, new traditions of vintage festivals and wineries begin to innovate new cellar doors

1966 World's most famous wine connoisseur visits

1968 Bushfires at Pokolbin destroy winery buildings

1969–75 Era of new corporate investment

1972 Inauguration of the Hunter Valley Vineyard Association

1973 Hunter Vintage Festival includes resurgent Upper Hunter subregion after contraction of planting there a half century earlier

By late 1970s Hunter wine companies at the forefront of Australia's new identity as a wine nation

Early 1980s shift in winemaking techniques to more specialist styles

1983–PRESENT – SECOND WAVE GLOBALISTS

HUNTER VALLEY HERITAGE CAIRNS	
Year conferred	Site
2009	Edward Tyrrell's Slab Hut – Tyrrell's Wines
2010	Old Cellar – Drayton's Family Wines & Old Vats at Audrey Wilkinson Wines
2011	The Old Still House – Ben Ean Winery & Maurice O'Shea Mount Pleasant Label – Mount Pleasant Winery
2012	Tulloch Wines, Pokolbin Dry Red Label
2013	Hall's Cottage – Roberts Restaurant circa 1876
2014	Marthaville Homestead & the 1973 Vintage Festival Poster, Cessnock
2017	Rothbury Cemetery

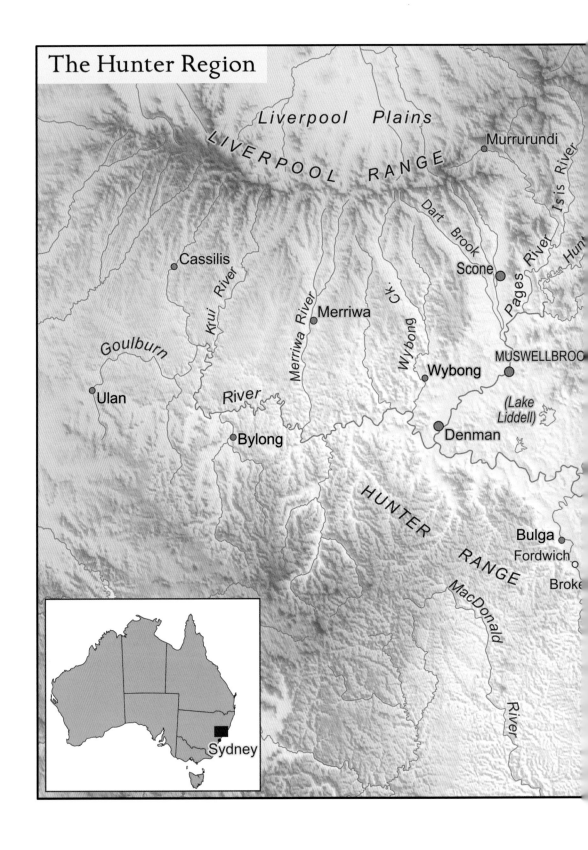

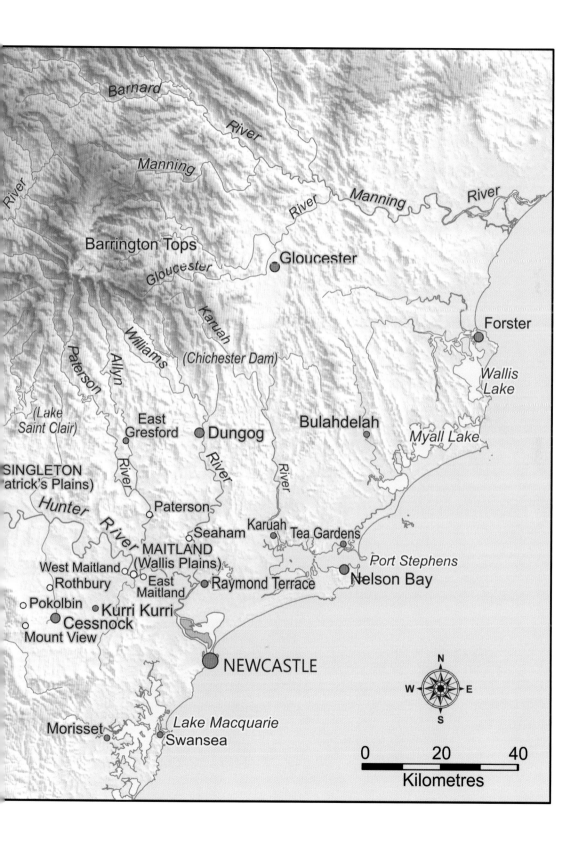

Introduction: making wine history

ON 10 JULY 1828, CHARLES FRASER – DIRECTOR OF THE Botanic Gardens at Sydney Cove – received a request for grapevine cuttings from wealthy landowner James Webber at Hunter's River, as the Hunter region of New South Wales was then known. Webber had sent his order for vines, on behalf of the Agricultural Club at Hunter's River, to Governor Ralph Darling. In response, the governor's assistant secretary, Thomas Cudbert Harington, wrote to Fraser asking that he send Webber 'as many Vine Cuttings in the Government Garden as can be spared [sic]'.[1] The cuttings were to be transported by coastal packet (ferry) from Sydney to the port of Newcastle, from where they would be transported upriver to Webber and his fellow settlers.

We have every reason to believe that Fraser shipped cuttings to Webber as requested, and that these were the stock for the initial vine plantings on a handful of properties throughout the Hunter – Australia's oldest continually producing wine region.

Geographically, the Hunter is a capacious, broad and long coastal lowland of close to three million hectares, nestled between the spurs of the Great Dividing Range. In 1828, when grapevines were freighted from the Botanic Gardens to Webber and his peers (located chiefly in the Paterson River valley, east of the main artery of the Hunter River), goods transport from Sydney to these settlers was by boat, and took nearly 24 hours. This journey proceeded along the coast to Newcastle by the *Lord Liverpool* packet, a former pleasure yacht from British India, and then upriver by smaller boat. Today, to drive by car from Sydney to the Lower Hunter's winegrowing heartland is a journey of less than two hours. A further hour of travel north-west will bring you to Upper Hunter wine country.

Pokolbin, c1949. Photograph by Max Dupain. Tulloch Family Collection, courtesy of J.Y. Tulloch.

Wine grapes were planted in the Hunter from 1828, and wine made there from at least 1832 by employees of the Australian Agricultural Company at Tahlee. The Australian Agricultural Company – the colony's first joint stock venture, established in 1824 – joined investors from Britain and New South Wales in an enterprise combining several land uses: grazing sheep for wool; crop farming; and horticulture, including wine grapes and olives. The company received a total of a million acres of New South Wales land, half of which, in the 1820s, lay between the Hunter coastline at Port Stephens and the south bank of the Manning River further north. Wool and shipping entrepreneur John Macarthur and his family members dominated the company's colonial board.

Landowner James King of Irrawang, 40 kilometres west-southwest of Tahlee, achieved the earliest recorded commercial export of wine from the Hunter, sent in barrels from Morpeth wharf on the Hunter River to Sydney in 1846.

Established at the birth of the modern global age of winegrowing at the turn of the 19th century, the Hunter is one of the most enduring New World wine regions. To put this in context, some mainland European winegrowing countries trace their antiquity back 2000 years to the Roman empire. In the early modern age, dating from the 15th century, Spain and Portugal established wine industries in their colonies in the northern Atlantic; the Portuguese in Madeira and the Spanish in the Canary Islands, each of which played a role in the colonisation of New South Wales. It's important to note that in wine-appreciation literature, wine industries with colonial histories are often termed New World, in contrast with mainland European wines, called Old World.

The Busby myth

THERE CAN BE NO CHALLENGE TO THE HUNTER'S STATUS AS AUSTRALIA'S oldest wine region on the basis of continuing production. Wine grapes long ago disappeared from production in the Sydney district where Australian winegrowing began. While some other Australian areas have appreciable histories of vine plantings and winemaking, the Hunter's continuity of winegrowing is unique. *Hunter Wine* is about the successive generations of vinegrowers, winemakers and wine business people who made this so. It is contentious, however, to credit the origins of Hunter vine plantings to Webber and his coterie of fellow large-scale landowners, and to identify 1828 as the year of original vine plantings in the Hunter region. In some other quarters, agriculturalist and author James Busby is viewed as the sole instigator of winegrowing in Australia, and has even been described as 'the pioneer viticulturist of the Hunter – possibly in seniority, certainly in achievement'.² Crediting Busby as *the* instigator of Australian winegrowing or *the* pioneer Hunter viticulturalist grew from a wine marketing campaign in the 1910s (discussed in chapter 7), and came to be compounded through a nationalistic tendency to seek out and applaud stories of singular pioneers of industry. Busby's name now appears in many places in conjunction with Australian wine, from books to wine labels to stories on the internet, but there is a history of colonial politics behind his elevation to this status – including a 'cold shoulder' to William Macarthur (son of John and Elizabeth Macarthur), who made a far more sustained contribution to Australian winegrowing. In the Hunter Valley, Busby's story – incorrectly told – overshadows a history of co-operation within families and between neighbours, beginning with the Agricultural Club at Hunter's River in 1828.

The evidence for dates and details of James Busby's actual contribution to winegrowing in Australia, and in the Hunter, arises in the most part from the Busby and Kelman family papers held at the State Library of New South Wales, and from combined research efforts on other documents.³ It is a strange and unsettling thing to dismantle a myth when the mistaken hero appears in his correspondence with his immediate family to have been a warm-hearted, if somewhat officious, character. There is, for instance, a line in a letter from James Busby to his (then) future brother-in-law William Kelman (who was integral to earning the Hunter a fine reputation as a wine region from the 1840s), in which Busby expressed displeasure at lack of attention from Governor Thomas Brisbane (Governor Darling's predecessor). Busby wrote to Kelman that Brisbane had treated him poorly, 'but I would not have this said again', he emphasised with an underline of his pen.⁴ When he wrote these words in 1824, it could not have occurred to Busby that over 190 years later his confidence to a friend would be read as part of the process of puzzling out his legacy in the emergence of Australian winegrowing. It is evident from the Busby and Kelman papers that Busby sought a career in the British imperial public service and that he hoped to make his mark as an important figure in the British colonies. (Indeed, Busby would later be appointed the first British Resident of New Zealand and oversee the drafting of the Treaty of Waitangi.) Busby's books and letters show that he decided that attention to winegrowing might provide him with fame and wealth. It is a shame that the shaping of Busby's contribution over the past century has distorted his story in ways that now need correction.

A brief history of early Australian winegrowing

HISTORIANS NOW LOCATE THE FORMATION OF THE AGRICULTURAL COLONY of New South Wales at the end of centuries of European encounters with Indigenous peoples throughout Australasia and Oceania.⁵ The origins of European Australia date from the establishment of New South Wales, at present-day Circular Quay in Sydney, by Governor Arthur Phillip in January 1788. The formation of the colony on the land of the Gadigal people of the Eora nation involved 11 shiploads of men, women and children, and many species of plants and animals foreign to Australia. Of about 1500 people who sailed with Phillip on the First Fleet from southern England in May 1787, bound for the east coast of Australia, about half had been convicted of petty crimes for which they had received sentences of transportation ranging from seven years to life. Most of the non-convict immigrants were soldiers of the garrison for the colony. Convicts were transported to New South Wales as labourers, to transform what British imperialists perceived as land insufficiently profited from by the Aboriginal peoples whom European adventurers had encountered at various places on the Australian coastline in the 17th and 18th centuries. In the 1780s, the British Crown and parliament sought to rebuild an empire substantially depleted through the independence of the former

American colonies, constituted as the United States in the 1770s. As no other European power had yet begun to profit from land in Australia through farming, grazing or mining – the traditional exploitative industries of European empires in new worlds – the British Colonial Office dispatched the First Fleet, constituting military garrison and convict labour force, to seed a new colony.

The First Fleet voyage took eight months. During this time, the ships of the fleet laid over for supplies, rest and repair at Tenerife in the Spanish colony of the Canary Islands for a week; in the Portuguese colony of Rio de Janeiro for a month; and at the Cape colony in southern Africa, then administered by the Dutch East India Company (the Verenigde Oost-Indische Compagnie, or VOC), also for a month. At these supply ports, Phillip made many official purchases for the new colony, including wine grape plants or cuttings (various cultivars of *Vitis vinifera*) at the Cape in December 1787. Although Tenerife in the Canaries was, like the Cape, a famous wine-producing country, there were no vine plants available there, and Rio de Janeiro did not have vineyards – the wine in that Portuguese colony came from Portugal.

As part of early cultivation at the colonial site near Circular Quay, vines from the Cape were planted by Phillip's gardener in the Government House kitchen garden near the present-day site of the Museum of Sydney. Some other Cape vines were later planted at Rose Hill (Parramatta). It seems a small quantity of wine was made from grapes from the Parramatta vines in 1792 and transported to London on Phillip's return to England in that year. Over the intervening years to 1824, several experiments in vine plantings and winemaking occurred in the colony, undertaken by a handful of colonists, principally William Macarthur and Gregory Blaxland. By 1824, Blaxland, Macarthur and some others remained dedicated to trialling cultivars of *V. vinifera* imported from different sources, and to producing small quantities of wine; and the Australian Agricultural Company too intended to begin winegrowing.[6]

From 1821, the level of transportation of convicted criminals increased, serving to remove criminals from England's shores and to provide labour for New South Wales' colonial enterprises. New Colonial Office policies on the privatisation of land seized from Aboriginal peoples, and the availability of convict labour, attracted many settler capitalists and fortune seekers to the colony, as discussed in chapter 2.

In 1824, at the crest of this new wave of immigration, James Busby arrived in Sydney with his parents, his sister Catherine, and some if not all of his brothers: George, William and Alexander (Ellick) Busby. James Busby's father, John Busby, had been appointed to the colony as a government hydrologist.

During the voyage to Sydney, Busby had drafted the manuscript for his first book, which he self-published in Sydney as *A Treatise on the Culture of the Vine and the Art of Making Wine; compiled from the works of Chaptal, and other French writers; and from the notes of the compiler, during a residence in some of the wine provinces of France* (1825). Busby stated that his interest in winegrowing grew from an 1821 British parliamentary report recommending this pursuit in those British colonies that possessed a suitable climate for growing grapes (as Britain's climate precluded wine grape production). While in residence in France, Busby appears to have lived primarily in the Bordeaux winegrowing region. Once in Sydney, Busby published the *Treatise* by subscription, although there were not many paying subscribers.

> In 1824, at the crest of a new wave of immigration, James Busby arrived in Sydney.

The *Treatise* did, however, find a small audience among a modest number of settlers seeking to emulate the traditions of wine grape vineyards in the hinterlands of the most famous trading port cities of the day – places such as Bordeaux, where British merchants had for centuries blended wines from local vinegrowers and sent their wares to London and other trading centres. Bordeaux wines were known in Britain as claret. British merchants also bought, blended and traded wines from vineyards in the hinterlands of Spanish Jerez de la Frontera in Andalusia, the home of wine known as sherry; and from Oporto in Portugal, where grapes from the Douro Valley were made into wine the British called port. Other famous wine places that were colonies of European wine countries and well-known to British colonists in Australia were Madeira, a Portuguese colony with a community of British traders in the city of Funchal – and source of a wine the British called Madeira; and the Canary Islands, producing wine that Shakespeare's Falstaff called Sack, as well as other wines known simply as Canary. At the Cape colony were some very fine wines named as Constantia, as well as ordinary styles enjoyed by the captains and officers of trade ships who called at this busy settlement – which from the 17th century was a vital supply port connecting Europe to the lucrative Spice Islands in the East Indies (present-day Indonesia) that were exploited commercially by the Dutch.

The centuries-old British wine trade tradition of naming wines for places, with no reference to the actual grape varieties in the wine, continues to influence wine names throughout the world. Some of the perplexing names for Australian wine discussed later in this book might make more sense when it is considered that a wine called Cape Madeira could be purchased at the Cape colony in the early 19th century. Some concerted effort to match wine names to constituent grape varieties began in the 1960s, but grape varieties as named sources of the flavours of wine did not begin to enter Australian (or indeed international) wine culture until the 1980s. It is not traditional to buy wine on the basis of grape variety, such as Semillon, Shiraz, Verdelho or Cabernet Sauvignon – and the Hunter wine community was at the forefront of this practice when it began. This is an intriguing part of the unfolding of Australian wine history within global as well as domestic wine trade.

But that is getting ahead of the narrative.

Back in 1824, as soon as James Busby arrived in New South Wales he began to petition Governor Brisbane for paid colonial employment. Although Brisbane offered him £100 a year to trial winegrowing on the government account, Busby held out for a higher income, which seemed to be possible with his appointment as manager and agricultural teacher at the Male Orphan School in present-day western Sydney. The salary for this role was to come from a share of produce of the school, an institution established to train boys and young men (abandoned colonial-born offspring of convicts) to be useful members of colonial society. Some time after March 1825 (the year is not certain), Busby planted a vineyard on school property – but he would not remain at the school long enough to make wine from this vineyard. Busby also *briefly* owned a Hunter property called Carrowbrook, but he did not farm this land, and may not have even visited the property.[7]

Mistaken memories about Busby's vines

In 1830, James Busby self-published (again by subscription – and this time there were more subscribers) *A Manual of Plain Directions for Planting and Cultivating Vineyards, and for Making Wine in New South Wales*.[8] In combination with this publication, Busby promoted the distribution of vine cuttings from the Botanic Gardens at Sydney – 50 free cuttings each, to any colonist who applied. This program of distribution proved very popular, with many dozens of settlers of the non-convict class receiving vine parcels, although Busby had aimed the *Manual* at smallhold farmers, most of whom were former convicts, or the offspring of convicts.

In New South Wales, former convicts were known as emancipists – whether they were rich or poor; and the children of convicts, or former convicts – whether rich or poor – bore what later came to be known as the convict stain. This differentiation between convicts and non-convicts created a difference in colonial social status related to migration history and colonial birth. It cut across income strata, complicating traditions of the British class system based on privilege by birth, income and levels of education. This was also in flux in the 19th century, because some people of reasonably humble birth earned tremendous wealth as merchants.

Despite Busby's hopes to induce poorer settlers to plant wine grapes and produce their own wine – as a measure of a hybrid sort of biblical and southern European civility (convivial drinking unconnected with violence and crime) – those settlers most interested in the *Manual* and the Botanic Gardens' vine distribution were the usual suspects in early New South Wales winegrowing: settler capitalists.

Moreover, receiving vines through programs such as the Busby plan did not automatically make a vineyard. Wealthy landowner George Wyndham, for example, recorded in 1831 that his vines from the Botanic Gardens did not shoot and grow, although he does not explain the cause of this failure – it may have been due to his own lack of experience, which Wyndham strove to correct.[9]

Between the appearance of the *Treatise* and the *Manual* a great change occurred in the colony, with the arrival of settler capitalists from England and Scotland, particularly in several of the distinctive districts of the wider Hunter region. It is possible – *but not proven* – that James Busby planted vines from the Botanic Gardens at his father's property of Kirkton in 1830. This goes to the heart of what is meant by planting. Did Busby send the vines to the property (then managed by Busby's brother-in-law William Kelman)? Did Busby dig the ground for the planting of the vines? This is unlikely, as convicts did most manual labour, and Busby appears not to have been a hands-on agriculturalist, more of a theorist and supervisor. Did Busby kneel on freshly dug soil at Kirkton and place vine cuttings in the ground and scoop the earth over to cover them? Possibly. Still, even if he did this, vines planted in 1830 were not the earliest grapevines in the Hunter region.

These statements are intended to undo some of the knots that have formed falsehoods in memories about James Busby and his involvement in New South Wales and Hunter winegrowing. Having said all of this, Busby does deserve credit for promoting the idea of winegrowing in 1830. His efforts greatly spurred interest in this form of cultivation for land use, when broader experimentation served to propel vine plantings as one of many options available to landholders. Busby's books are important artefacts of wine history, but true stories of the past cannot rely only on selective use of some surviving records, and the Busby and Kelman papers introduce nuances to the idea of this one man as a sole pioneer.

In 1831, Busby returned to England for reasons related to his removal from employment at the Orphan School (not due to any misdemeanour on his part). While in Europe, Busby also travelled to Spain and France. There he collected many hundreds of vine cuttings to further encourage grape growing, a journey subsequently recorded as *Journal of a Tour through Some of the Vineyards of Spain and France* (1833).[10] Busby negotiated a subsidy from the Colonial Office in London to meet the costs of the freight of the vines that he donated to the Botanic Gardens in Sydney. Busby also distributed

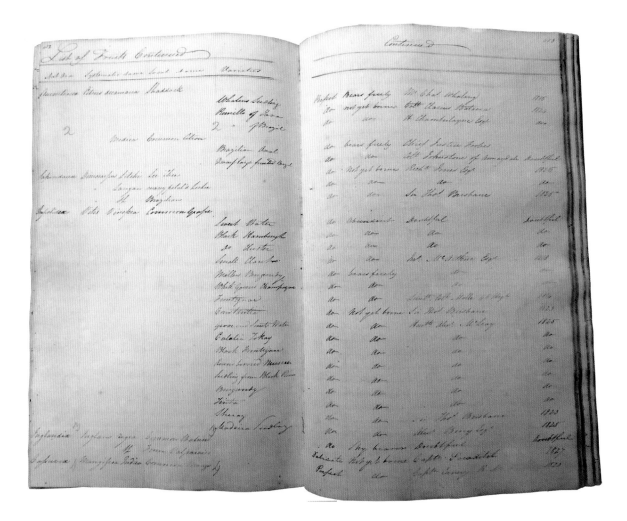

vines among his family members, some of which were sent to the Hunter. The distribution of these vines is also a part of the not entirely well understood Busby legacy, and the Busby vines from Europe are discussed further in chapter 2.

Historical errors and overconcentration on Busby have obscured many other significant characters and elements of Hunter wine history, especially the co-operative, rather than individualistic, nature of the enterprise. Our attention to many growers who conducted experiments from the 1820s – and the extent to which vine stocks were swapped and sold for over a century (in and out of the region) – emphasises that the

Vitis vinifera cultivars in the Botanic Gardens Sydney. List of fruits cultivated in the Botanic Gardens, Sydney, up to Nov. 1827, B1, p, 512, The Daniel Solander Library, Royal Botanic Gardens Sydney.

dominant story in Hunter winegrowing is one of co-operation and exchange, rather than individuals operating in a vacuum.

Mythmaking in the 20th century

IT IS NECESSARY TO NOW JUMP AHEAD A CENTURY TO FINISH CLEARING the air about Busby and Australian wine. In 1931, a federal government report into the Australian wine industry made this statement on its first page: 'In 1830 William Macarthur produced wine at Camden, and in the same year James Busby planted vines at Kirkton in the Hunter Valley'.[11] Busby *may* have sent vines to Kirkton, but he neither managed these vines nor made wine from them. In 1952, a Hunter newspaper mistakenly claimed that James Busby planted vines he imported from Europe at Kirkton in 1830. This is certainly untrue.[12] Leafing through many records, such misapprehensions became common – pointing to the tangled web of tales about Busby inadvertently woven through a combination of wine marketing and other sources.

The 1931 report also became the source of a continuing narrative about the unsuitability of the Hunter wine region for wine grape production.[13] The report's authors based this conclusion on the frequent occurrence of hail, frost and flood, and also on the lack of interest in Australia and Britain (a principal export destination) in dry table wines.

To explain the concept of dry table wines, as will recur later in the book: these are non-sweet wines, with a relatively low level of alcohol, achieved from fermentation of the fruit sugar of grapes into wine. This is in comparison with so-called fortified wines (fortified with brandy) – which have a high level of proof (percentage of pure alcohol per volume of liquid). In Australia, naturally occurring alcohol is high in wines from warm districts, such as southern New South Wales and South Australia. Hunter climate conditions historically produced relatively light or lower alcohol wines. Winemakers may predict alcohol volumes measuring degrees of Brix (sugar mass in grape juice) with a hydrometer. One per cent of grape sugar equals 1.8 Brix. Adding brandy to wine to increase alcohol levels came to Australia as part of habits of British wine trade practices.

Spirits stabilised wine, preventing a second fermentation in barrels during shipping; they also gave a kick to the taste and heady effects of wine, which became popular among some drinkers. Most Australian winegrowers of the 19th century expecting to export wine had to consider adding spirits as a stabilising agent. When some domestic consumption of wine began to arise in the 1860s, many Hunter producers preferred to make what they called pure wines, later referred to as non-fortified wines. Pure wines might be either ordinary or fine.

More on this in later chapters – the principal matter is that in the 1920s most Hunter wines were dry table wines. This meant that if they did have spirits added (and many did not), the alcohol content of the wines was below 33 per cent, as a fortified wine was defined as having one-third of added brandy. It is astonishing, therefore, to read the 1931 report's statement that as the demand for table wines (for which the Hunter had been famous for three generations) remained negligible, 'it appears doubtful whether this area can continue in production'.[14]

The 1931 report on the Australian wine industry, the first such inquiry to be held in Australia – and at a time when the wine industry was in crisis for other reasons (see chapter 7) – not only sentenced the Hunter to doom because of environmental challenges, but also predicted that a lack of interest in dry wines meant the Hunter was *doubly* doomed. The number of vineyards and wineries in the Hunter did shrink during the 1940s and '50s. When the tide turned in Australia's thirst for wines from the 1960s however, persisting Hunter winegrowers were joined by other landowners returning to the industry, and by new investors. This brought a wave of growth in tandem with social conditions conducive to creating a wine-drinking country.

Particulars of Letters received & answered in the Botanical Department of Sydney commencing 19 May 1828, A1, The Daniel Solander Library, Royal Botanic Gardens Sydney.

Grapes for wine

IN THE MIDSUMMER OF 1827, A SYDNEY NEWSPAPER PUBLISHED A HIGHLY ranked colonial gentleman's impressions of the Hunter, made during a journey he had undertaken earlier that season to the principal settler properties of the region's districts. These impressions appeared under the

Sydney 14th June 1828

Sir,

I have the honor to inform you that His Excellency the Governor has approved of the undermentioned Requisitions submitted by you; and that the several Articles specified therein have been ordered to be supplied by the Deputy Commissary General.

No. 1759 — 4th June, Garden Hoes, &c
    1761 — 8th June, Blankets. —

I am
Sir
Your obedient Servant
for the Colonial Secretary
T. C. Harington

The Colonial Botanist

Colonial Secretary's Office
Sydney 10 July 1828.

Sir,

In compliance with the request of Mr Webber, the President of the Farmer's Club at Hunter's River, I am directed by His Excellency the Governor, to desire that you will transmit by the Lord Liverpool Packet to that Gentleman, for distribution, as many Vine Cuttings in the Government Garden as can be spared.

I have the honor to be
Sir
Your most obedient servant

Mr C. Fraser

pen name 'XYZ', as gentlemen of this era contributed their thoughts to newspapers from behind a mask of anonymity. Indeed, the identity of XYZ still remains a matter of speculation, although his name is widely given as William Dumaresq.[15] There were no wine grape vineyards in the Hunter in 1827 – because if there were, Dumaresq would have wanted to see them. As he declaimed, 'no one likes grape more than John Bull' (that is, middle class British capitalists such as himself), 'yet without a single exception, not more, I never saw in New South Wales a bunch worth eating. Wine is therefore out of the question, at present'. Through encouragement of the cultivation of 'plenty of grapes', however, Dumaresq expected that 'wine will follow as a matter of course', in places such as the Hunter.[16]

Visiting the Hunter, Dumaresq travelled the 50 or so kilometres from Newcastle to Wallis Plains (Maitland) by riverboat. Of the landscape in this area he asked 'why a low flooded forest should be called Plains?'. These 'plains' have long since been drained of wetlands. After leaving the river, Dumaresq soon encountered what he considered (with typical British imperial hyperbole) to be 'one of the finest countries on earth'; perfect for agricultural development.[17] He decided that the 'rich alluvial flats' near Maitland, once cleared of timber, would 'repay the settler with rich and abundant harvests'.[18]

From Maitland, Dumaresq rode north-northeast through the bush (he called this 'forest') to 'the Old Branch or Paterson Plains', where he very likely met with Webber and others from the Hunter's River Agricultural Club, whose membership is discussed in chapter 2. In the Paterson district, the travelling gentleman perceived 'a change for the better ... the land being equally good, and a large space of country on each side of the river cleared and under cultivation'. Moving on from Paterson, south-east to the Williams River district near present-day Raymond Terrace, Dumaresq mused that this area might answer well for colonial improvement but was hampered by drunken lower class settlers, and that there one slept with 'pistols under the pillow'.[19]

Turning westward towards Patrick's Plains, or contemporary Singleton, Dumaresq's journey took him and his party through 'the dark and murky forest' until, suddenly, under a 'yellow moon', the travellers 'burst upon

Patrick's Plains, the first sight of which, and the sensations it awakened, I shall not forget the longest day I live. I was completely enamoured with the splendid scene ... an almost boundless level without a tree'; a place, decided the gentleman, that 'deserves the name of Plains'.[20]

The cleared country at Paterson and the splendid treelessness of Patrick's Plains were the work of Aboriginal land managers before colonisation. Dumaresq almost certainly undertook his journey guided by at least one Aboriginal man, or boy, as colonists rarely ventured into the bush in the Hunter without a local guide until as late as the 1840s. Unfortunately, Dumaresq does not name his guide or guides, which makes it seem as though this assistance did not exist; yet strong bonds of reliance were crucial between settlers and Aboriginal people in this period in the Hunter.

Each of the places visited by Dumaresq became significant in early Hunter winegrowing because settlers were clustered there. Europeans had also settled in the Upper Hunter and at the mouth of the Karuah River near contemporary Port Stephens, centred on the Australian Agricultural Company's property Tahlee, a headquarters for early timber-getting as well as a farm. In 1827, a few months after Dumaresq's Hunter encounters, James Macarthur – a second generation settler and founding member of the Australian Agricultural Company with his father John, and others – visited Tahlee. Following this visit, James reported that the property's kitchen garden contained *grapevines* as well as fruit trees.[21]

> Arriving at Patrick's Plains, Dumaresq 'was completely enamoured with the splendid scene ... an almost boundless level without a tree'.

Perhaps Tahlee's vines, rather than those requested by Webber in July 1828, were the first wine grapes planted in the Hunter? If we continued to search, would we find more evidence about the timing of the planting of these Tahlee vines? We think not. Following on from his son's visit, John Macarthur called at Tahlee in May 1828 accompanied by Fraser, the director of the Botanic Gardens. Macarthur sought Fraser's advice on a suitable location for a 10 acre vineyard, for which Macarthur proposed to provide vines cultivated by his son William at the family estate at Parramatta.[22]

Could vine cuttings have been sent from the Botanic Gardens to Tahlee later in 1828? Certainly. And not to Webber? This seems unlikely. What is clear, however, is that the Hunter's first generation of settler winegrowers were part of a wider determination to acclimatise *V. vinifera* in the region, and that the Hunter offered great promise as a place for wine production.

In the early days of the new British colony, wine grapes were a small but significant part of a vast complex of cultivations envisaged by the architects of the New South Wales enterprise in the late 1780s.

At first only Webber and a few of his peers planted vines in the Hunter – because they were wealthier than most other settlers and wine grapes

In the vicinity of Tahlee. A native camp near Port Stevens, New South Wales. Augustus Earle, c1826. National Library of Australia, Rex Nan Kivell Collection, NK12/28.

are a costly form of cultivation. Compared with annual grain crops and small-scale grazing of livestock on native grasses (common pursuits in the Hunter's early settler history), vineyards were expensive to plant and slow to show a profit. That was, until the 1860s, when land policies allowed for more smallholders – who planted grapes to sell to wineries, or otherwise benefited from the long processes of experimentation begun by others.

Early Hunter landholders were primarily British, and did not have a homeland culture of growing grapes for wine – in contrast with, say, some later German immigrants, and those Italians arriving in postwar Australia in the 20th century who grew grapevines alongside tomatoes and greens in their backyards in Newcastle, Sydney and Melbourne. Among Hunter settlers, early wine production was at the cottage scale, but evidence indicates these settlers looked to a further horizon of commercial production: they sought to make wines that could take their place within the colonial wine trade, which was dominated from 1788 to the 1870s by imported wines.

Some wine styles made in the Hunter today are the result of long association with specific grape varietals: Semillon, Shiraz and Verdelho. Certainly, Shiraz or Red Hermitage from Kirkton proved successful in the 19th century. There is no way of telling precisely which grape varieties were first planted in the Hunter in 1828, but at that time the Botanic Gardens in Sydney held 16 cultivars of *V. vinifera*; many were bearing sufficiently to provide canes from prunings as clonal plant stock, and most were wine grapes. The available varieties were:

- Sweet Water (an eating grape)
- Black Hamburgh (likely a garden grape from temperate England)
- Black Cluster (a garden grape from south-west England)
- Small claret (perhaps Cabernet Sauvignon)
- Millers Burgundy (perhaps Pinot Meunier)
- White Gouais Champagne (Gouais was widespread in European vineyards)
- Frontignac (a Muscat, likely from the Cape)

- Constantia (definitely from the Cape, speculatively Semillon)
- Grave (perhaps Semillon or Sauvignon Blanc) and Sweetwater (how this varies from the earlier is not clear)
- Calalia Tokay (from Hungary's Tokaji industry)
- Black Frontignac (Muscat)
- Round-berried Muscat
- Seedling from Black Prince (Cinsaut)
- Burgundy (perhaps Pinot Noir)
- Tinta (often used in port wines), and
- Shiraz (a French grape, known too as Red Hermitage).[23]

The original stocks of many of these varieties were donated to the Botanic Gardens over the course of a decade by, variously, Governor Brisbane (the Shiraz), a governor's secretary and John Macarthur; some others came from unidentified sources. (Importations of vine stock were regulated from the 1860s to prevent the spread of vine disease, and later vine pests. Before then documented imports were not the only vine stocks in Australia, but knowledge of vine types is limited to those records that do exist.)

Today, vine plantings across the totality of the Hunter region are centred in six subregions: the Upper Hunter near Denman; and in the Lower Hunter – Pokolbin, Broke-Fordwich, Wollombi, Lovedale and Mount View. According to geographer Phillip O'Neill, the Hunter region lacks soil types that some other Australian winegrowing regions promote as ideal, and possesses instead 'undistinguished heavy clays'.[24] Historically Hunter vineyards have been planted in soils that vary in type and contemporary soil mapping shows the Hunter's soil profile is diverse. What distinguishes the Hunter is that, although the region possesses sizeable portions of land ideal for agriculture, the terrain is not conducive to large-scale viticulture (grape growing), nor to broadacre farming of grain crops. Wine production volumes from Hunter grapes are traditionally modest, focused more on quality than quantity.

While there is no doubt that the vigour of grapes derives from their soils and the wider locale, scientists emphasise that soils may be readily managed to achieve an 'expression of place' through encouragement of humus organisms, application of fertilisers, and the quantity of water that vines receive naturally or through irrigation systems.[25] In a contemporary world of thousands of wines from dozens of countries, the idea of special soils and natural environments is a very useful way to distinguish some wines from others. This notion of the physical environment of wine grapes as the most important factor in wine quality is called in the English-language wine trade *terroir*, the French word for place of origin – and is associated historically with the industry narrative about wines from Burgundy in France.[26] The concept of *terroir* used this way (rather than inclusive of the 'hand of man', as sometimes occurs) is, in actuality, dehumanising. To invest soils and other elements of non-human nature with a sense of divinity has a cascading effect that ultimately stifles vigorous historical knowledge of a wine region. First, it obscures the wider ecology of winegrowing as an enthralling product of human ingenuity that occurs in a natural environment containing more than grapevines.[27] Second, history is contingent on the presence of people, as agents of continuity and change. Without people in the story, vineyards and wine make no sense.

> What distinguishes the Hunter is that the terrain is not conducive to large-scale viticulture. Wine production volumes are traditionally modest, focused more on quality than quantity.

As geographer Brian Sommers points out, the wave of wine globalisation since the 1990s has led to a great levelling in access to wine and to an astounding expansion in the places where vineyards are planted for wine. Under these circumstances, attention to the distinguishing physical features of wine regions is a way to unlock understanding of the exceptional qualities of winegrowing places as spaces with meaning.[28] Regional (and intra-regional) distinctions may be read in the appearance of vineyards (height of vines, spacing of rows), specialisations in grape varieties and blends, positioning of vineyards relative to natural features such as soil types, landscape contours, watercourses – and (environmental humanists would add) co-existing plants and animals. Out of these factors the one that should not be forgotten is the role of the people who select vineyard sites, those who nurture grapevines, those who choreograph the

annual grape harvest and are otherwise engaged in the risky artistic alchemy – and the scientific precision – of making wine.

New wine regions may be designed on the basis of suitability of location, but for the Hunter, winegrowing is entrenched, and the history of the people who have made their livelihoods there, and the wines they have made, created an imperative for perseverance. When Daniel Honan conducted an environmental history project to ask why Hunter winegrowers persist in the region – if the conditions are purportedly less than ideal – he found that most growers stay because the Hunter is their home, and they have a continuing fascination with the characteristics of their wines.[29] We are not claiming universal harmony among Hunter winegrowers, yet the persistence of contemporary growers is linked to the strong history of a circuitry of connections between many people in the region. These connections have been a distinctive and critical feature of the Hunter winegrowing community since Webber and his peers planted their vines and visited neighbours like Wyndham, who also planted vines ... and so on.[30]

There have been seven settler generations engaged in winegrowing in the Hunter. We have identified and named them the imaginers (1828–46), the experimenters (1847–76), the first wave globalists (1877–1900), the forgotten generation (1901–30), the stoics (1931–54), the Renaissance generation (1955–83) and the second wave globalists (1984 to the present). In succeeding chapters, we first examine the preconditions for Hunter winegrowing, and then map out the first six of these generations – leaving future historians to tell the story of the seventh. We devote a chapter to people from the south-west German region of the Rheingau (the Rhine district, as it is known in English) who immigrated from the 1840s as labourers, and some of whom became winegrowers.

A note on Hunter wine historiography

THE HUNTER WINE COMMUNITY'S LONG HISTORY IS LITTLE UNDERSTOOD within the community itself, within the wider region, or further afield. Hunter wine, and the region broadly, has low visibility in national history. There are four reasons that winegrowing, trade and drinking culture have

PREVIOUS PAGES: Mount Pleasant winery. Photograph by Max Dupain, c1950. Tulloch Family Collection, courtesy of J.Y. Tulloch.

been of little interest to Australian researchers until the past decade. First, wine has historically been conflated within the study of alcohol purely as the basis of social and health disorders. Australian history as it is taught in schools and universities began to be written in the 1950s, at the tail end of the temperance era (see chapter 7). Due to disapproval of alcohol among the educated and political class of the mid-century, historians overlooked the robust ties of many key colonists such as Macarthur and Blaxland to winegrowing. Second, wine had little economic significance in Australia, compared with some other export commodities, until the 1990s. And, once the Hunter wine region became part of a national wine industry, with the Federation of the Australian states in 1901, the natural limits on Hunter wine production (compared with some other regions) excluded it from a national economic paradigm of continual expansion and growth. Third and fourth, agricultural history (of which vine cultivation is part) has negligible status in Australian economic, social, political and cultural history – and, until recently, winegrowing has had marginal status in the field of environmental history.[31]

In 1969, Newcastle historian WP Driscoll considered the development of Hunter winegrowing from 1832 to 1850. Driscoll named some of the early key figures, mapped some of the properties where experiments occurred, concluded that the Hunter benefited from the timing of its origins within the colonial economy but that winegrowing remained nascent in 1850, and recommended further study.[32] Since the 1990s, historians and sociologists have joined geographers and anthropologists in paying attention to the role of wine in national and global history and contemporary society.[33] Yet Hunter wine history has received little attention until now within this burgeoning literature. Besides Driscoll's research, we have written on the early years of vine and wine production at Dalwood, and concluded that a long history of family and social ties has been a source of contemporary resilience in the Hunter wine industry community.[34] We have also shown how the tremendous popularity of Lindeman's Ben Ean Moselle – a wine invented in the Hunter – gave many Australians a new taste for wine in the 1970s, during that decade's extraordinary shifts in gender, class and cultural nationalism.[35]

We have found that some people suspect 'wine history' of having impure economic motives, compared with some other forms of historical research. Some years ago, an episode of the ABC's national radio program *Bush Telegraph* was centred on our concern that a Hunter wine heritage site – Dalwood at Wyndham Estate – was to be closed to the public.[36] During the program, a caller remarked that Wyndham Estate could only be important as a wine brand – not as national heritage – and that it was immaterial whether people could visit the property or not. The caller's comment draws attention to the wine industry's selective use of the past to sell products, but this is not the only possible permutation of 'wine history'. The Hunter is unique in Australia for the longevity and durability of its vine plantings and wine production. This is worthy of historical sociological study across the sweep of continuous winegrowing generations, to understand the enduring relationship between a community, a product and a region. Some brands inevitably appear in our history, but we are also engaged with Aboriginal–settler relations, the ethnic diversity of settlers who engaged in winegrowing, the politics of imperial land use, the emergence of colonial business, and science and technology.

> The Hunter is unique in Australia for the longevity and durability of its vine plantings and wine production.

Our primary sources for this book include public and private archive collections, ranging from correspondence to business papers, images, transcripts of existing oral history recordings, new oral histories, newspaper articles, television advertisements and film. We have crisscrossed the Hunter Valley by car, tramped across paddocks, and visited churches, graveyards, vineyards and homes. We have been advised by local and family historians, and have spoken with many living members of the Hunter wine community and with relatives of members of past generations who have been forgotten.

In drawing on a diversity of documents and material culture, we have set out to understand the emergence of Hunter winegrowing as an 'entangled history' of threads of inquiry.[37] Our investigation of the emergence of Hunter wine is a new lens on winegrowing as a form of specialist horticulture anchored through vineyards and wine business ownership to locale – but also bound to wine traders and drinkers outside the region. Over time, the concurrent combinations of winegrowing people in neighbourly co-operation and competition within the Hunter have also been involved in (as well as influenced by) policy and trade at national and global scales – so that the community has in turn become a motor for changing laws, practices and rituals.[38] Hunter wine began from settler desire to create a new form of cultivation in a new colonial territory, and across subsequent generations the continuation of this ambition has changed the Hunter region and its people, and broader fashions for wine.

Chapter 1

Vitis vinifera in Aboriginal Country: introducing an ancient plant to an ancient land

MAKING WINE FROM THE FRUIT OF *VITIS VINIFERA* IS thought to have originated in Georgia of eastern Europe, some 8000 years ago.[1] Even so, understanding Hunter traditions of winegrowing and wine business requires attention to where *V. vinifera* was introduced, as well as to where it came from. The first thing to be said is that all Australian plantings of *V. vinifera* are made on land once occupied and cherished by Aboriginal locals. These locals likely created the park-like clearings in the Hunter landscape where some of the earliest vineyards were planted. Moreover, Aboriginal people lived for at least a generation on colonial estates associated with early Hunter winegrowing, such as Charles Boydell's Camyr Allyn on the Paterson River near Maitland, the Scott brothers' Glendon near Singleton, and George Wyndham's Dalwood on the Hunter River near Branxton.

The invasion of Aboriginal lands in the Hunter began in the late 1790s, when Aboriginal guides and military guards pursued a group of convicts along the coast northward from Sydney. Although the precise duration of the Hunter's human history is not known, Indigenous Australian habitation is now dated back some 65000 years, and this ancient occupation came after a vast expanse of geological time.[2]

Before human history in the Hunter

THE CONTINENT OF AUSTRALIA POSSESSES SOME OF THE WORLD'S OLDEST recorded rock formations, yet the physical geography of the Hunter region – actually not one valley, but a cluster of valleys centred on the Hunter River – is relatively young.[3] The Hunter's dominant geomorphology began approximately 380 million years ago. During this most recent 8 per cent of planetary history, glaciers have formed and melted, volcanoes erupted and cooled, and seawaters encroached and receded within the extensive Sydney Basin – of which the Hunter forms the north-eastern edge.

The Hunter's geography is divided by its tectonic feature, the Hunter-Mooki Fault Thrust system, which separates the north-eastern portion of the Williams and Paterson river valleys (to the Upper Allyn River and Barrington Tops) from the remainder of the region. The difference can be seen when driving around this part of the valley, as the Paterson and Gresford district is hillier than the plains areas. It also features naturally occurring vegetation that is, in parts, more lush and green than the blue-green eucalypts and kangaroo-hide coloured native grasses of other parts of the Hunter region. The geology of the north-eastern portion – the Paterson through to the Upper Allyn – is Carboniferous, made from rocks formed to 360 million to 300 million years ago, compared with the later, more readily eroded, Permian rocks of the central lowlands that are 299 to 251 million years old.

After the formation of these rocks, across hundreds of millennia, many forces of non-human nature changed the Hunter region's coastline, uplifted escarpments and shaped the river system.[4] Excluding the stretch of coastline from Lake Macquarie to Port Stephens that opens the wider valley to the sea, the Hunter region is encircled by spurs of the Great Dividing Range. These uplands direct the flow of water seawards, into the broad, shallow bowl of the lowlands. This gravitational sedimentary flow produced the region's principal watercourses, as rivers rising from upland wellsprings once charged almost directly across the plains, bringing the sediments that are the building blocks of soils. The earliest flowing waters deposited rich muds, composed of mineral sediment and organic matter, on the banks of waterways, thereby sculpting serpentine riverine meanders.

The region's watercourses are the eponymous and arterial Hunter River (known to Aboriginal people as Coquun, and before the 1840s as Hunter's River), the lower Hunter River branches of the Williams and Paterson rivers, called Dooribang and Yimmang before colonisation, the Goulburn River in the western portion of the Upper Hunter region, Wollombi Brook in the south-east of the Lower Hunter, and the Karuah River that flows into Port Stephens at the north-eastern border of the region.

The northernmost Hunter Valley township is Murrurundi (which means 'five fingers', named for the rock formations of the district), lying east-northeast of Cassilis. Northward beyond Cassilis and Murrurundi are the rich black soils of the Liverpool Plains in the New England region – ideal for cultivating grains and pasture crops. With the exception of the (black to brown) basalt soils of the low-rainfall Merriwa district to the east of Cassilis, most Upper and Lower Hunter soil types are broadly defined as texture-contrast. This means that the type of topsoil varies from clay to sand. These combinations of soils were formed from differing layers of mineral sediments and organic matter that accumulated over millions of years, and which were then disrupted by earthquakes. Coal deposits below the Hunter's topsoils were evident to early colonists as a result of the texture-contrast exposure of under-earth layers (and the presence of coal seams continues to attract mining investment to the region). The texture-contrast characteristics of the Hunter also explain why some areas of the region located away from the river-bank alluvium (rich organic soils) have red clay, or white sand or, more rarely, black soils. Theoretically, the Hunter offered colonists many acres of suitable vinegrowing land, but other factors decided where vineyards would be planted – such as the ideological and economic disposition of certain settlers towards winegrowing, as discussed below.

Over millions of years, the combined action of sedimentary flow and earth movements created the conditions for certain types of plant life. In the Upper Allyn and Barringtons, they fostered wet eucalypt forest and rainforest – large trees with dense plant understoreys – conditions

Thought to be Camyr Allyn. Conrad Martens, 'On the Paterson River', Album of cloud studies, mountain, bush and harbour scenes, c1841–50, State Library of NSW, Call no. DL PX 28, frame 13.

ill-suited to grapevines, which require abundant sunlight and air circulation as well as soils that do not hold too much moisture. The native vegetation of the remainder of the Carboniferous edge of the region, and the Permian midsection, historically hosted ironbark trees and box gums standing in native grasslands, with fewer ironbarks north of Muswellbrook to the Liverpool Range. The Cessnock and Wollombi districts of the Hunter – away from the alluvial flood plains and eucalypt woodlands of the river junction of the Hunter, Paterson and Williams rivers – feature a combination of these woodland types, with a grassy or shrub understorey, and dry eucalypt forests.[5] These dry forests represent the quintessential Australian 'bush', characterised by sparser foliage and less-frequent rainfall. Some of these differences can be seen between lusher vegetation in the Lovedale and Mount View vineyard subregions, compared with Broke, which is in dry eucalypt country.

First peoples of the Hunter

THE EVOLUTION OF PLANT AND ANIMAL LIFE IN THE VARIOUS RIVER valleys of the Hunter region shaped Aboriginal traditions of finding food, water and shelter, and through managing these resources Aboriginal people developed their spiritual rituals and symbols of being. The district of Wollombi, located between Mount Yengo, at the southern border of the Hunter region, and Wollombi Brook, south-west of Pokolbin, is a known locality of Aboriginal culture. Many ancient artworks and walking tracks in the Wollombi Valley suggest habitual gatherings by the ancestors of Aboriginal communities who identify as the Awabakal (of Newcastle), the Wonnarua (of the Singleton district), the Darkinjung (from the Upper Hunter to the Hawkesbury), and the Dharug (from Windsor near Sydney). According to Aboriginal community leader Bill Hicks, Wollombi means 'a meeting place, a meeting of waters or meeting of peoples'.[6] Research also points to ancient and continuous Aboriginal occupation in other parts of the Hunter Valley prior to colonisation, including the Worimi people in the Karuah River district.[7]

While fire-farming is understood by many historians to be a part of Aboriginal land management, Bill Gammage used colonial documents and images of Australian woodland regions to piece together a picture of fire-farming practices by Aboriginal people dispersed widely across the continent. 'What fire to use and when varied', argues Gammage, 'but the purpose was the same, to associate water, grass and forest, providing habitats [for native animal prey] and making the clean, beautiful landscapes dear to Aboriginal feeling'.[8] Gammage drew some of his evidence from the Hunter, such as Joseph Lycett's vibrant colonial paintings of Aboriginal people and landscape in the Paterson Valley.[9] Moreover, areas of the Paterson and other parts of the Hunter were observed by early colonists to have wide grasslands (although overgrazing by sheep and cattle destroyed these within a generation of settlement).

Historian Grace Karskens, who studies Aboriginal and settler lives in the Sydney region, argues that Aboriginal people also lived in places like river corridors and rainforests with plants that had not evolved to respond to fire.[10] This perspective emphasises the importance of viewing Aboriginal practices at the district level (or local, rather than national scale) as connected to actual ecologies.

In 1826, settler Peter Cunningham observed the 'luxuriant plains' of the Upper Hunter to have 'scarcely a superfluous tree' – not more than a dozen eucalypts per acre, which eased the effort needed for settlers to plough the land ready for crops.[11] This landscape was very likely the result of Aboriginal control of understorey growth through fire. As discussed in the Introduction, when gentleman traveller and writer XYZ – William Dumaresq – encountered Patrick's Plains (near Singleton) in late 1826 or early 1827, he marvelled at the extent of grasslands in that district in a way that pointed to existing Aboriginal management. From Dumaresq we learn too that locals called the 'pretty' Paterson Valley 'Yua Lunga', and 'have a name for every nook and corner of the country – every variety in the vegetable kingdom – including the smallest flies and ephemera of the air'.[12] Cunningham described how even in dry years when the usual ('delicious') fish stocks were low in Hunter's River, Aboriginal locals harvested '*cartloads*' (Cunningham's emphasis) of 'immense eels' from muddy pools; and he also commented that on Patrick's Plains, majestic wild turkeys were 'an excellent and most delicate repast'.[13]

A guide for immigrants to New South Wales published in 1826 lists the native species of trees prospering within the Hunter region as 'Rose Wood', 'Cedar' and 'Coal River Pine', 'Blue Gum', 'Blackbutt Gum', 'Flooded or Water Gum', 'Iron Bark', 'Stringy Bark', 'Turpentine', 'Mahogany' and 'Sassafras or Kalang'.[14] This combination of English with some Aboriginal names signifies an enmeshment of British and Aboriginal cultures as colonists came to grips with Australian nature, often in shared journeys with Aboriginal locals who were intimately acquainted with the landscape, and without whom settlers were helpless in the unfamiliar bush.[15]

Manifestly, the Hunter was shaped and valued by Aboriginal people in ways that benefited colonists once they began to arrive in the region from 1797, and this sets the scene for how closely locals and settlers were entangled on the early colonial frontier.

Connecting the Hunter to the colonial economy

IN 1797, CONVICT ESCAPEES STOLE THE *CUMBERLAND*, THE VESSEL generally used for ferrying goods and people between Sydney township and the Hawkesbury settlement. In searching for these pirates (and this valuable boat), Aboriginal trackers (their names are not recorded) led Lieutenant John Shortland of the colonial garrison to Coquun. Shortland renamed it 'Hunter's River' for acting Governor John Hunter – before proceeding to Port Stephens, about 200 kilometres north of Sydney's Port Jackson. In the southern escarpment at the mouth of Hunter's River – and perhaps on the river shore – Shortland spied visible deposits of coal. Here, and along the coast, Shortland also noted stands of red cedar (*Toona ciliata*). Red cedar produces a magnificent mahogany-like timber from a soft, easily carved and spicy-aromatic central core or heartwood, which ranges from pink to red in colour. These trees were prized in colonial times for making elegant household fittings and furniture. Historian Tamsin O'Connor considers Shortland's 'discovery' of coal and cedar to have been 'consolation' for the loss of the *Cumberland* and the pirates, who remained at large despite the search.[16]

To extract this coal and cedar, a new colonial station was established in 1804 at the mouth of the river at King's Town – later named Coal River, and then Newcastle. Colonial officials soon designated this as the first site for internal transportation as a punishment for crimes committed elsewhere in the colony. At Newcastle, convicts mined coal, cut and dressed timber, and burned lime for mortar using the shells from middens left by many generations of Aboriginal people feasting on oysters from estuarine coves. The non-Aboriginal population of Newcastle was nevertheless very small, and resource exploitation languished until the arrival in 1809 of Governor Lachlan Macquarie. The new governor's vigorous program of public construction and private investment provided an impetus for Newcastle industries. Furthermore, between 1815 and 1821, the British colonial office greatly increased the number of convicts shipped to New South Wales, enhancing Newcastle's threefold role as a provider of colonial resources, a destination for colonial offenders and an employer of new transportees.[17] During this period, the river port of Newcastle played an important role as the entrepot for water transport between Sydney and burgeoning Hunter hinterland settlements. Convict escapees at large came into contact and conflict with Aboriginal locals, a part of the story of early settlement that is hidden from view when focusing only on economic development. These encounters taking place behind the scenes during the advance of settlers upriver in the Hunter Valley form a crucial backdrop to later recorded cases of frontier violence in the hinterland.

In 1812, as part of his energetic administration of New South Wales, Governor Macquarie fostered small-scale farming at Paterson's Plains (which by the 1840s would become known as Paterson Plains), north-east of Maitland in the Lower Hunter, a few kilometres from the present day village of Paterson. In that year, as a reward for their outstanding work as timber sawyers, Macquarie provided conditional pardons to convicts Benjamin Davis, George Pell, John Reynolds and John Swan, and allocated them small acreages of land. A fifth sawyer, John Tucker junior, also settled near Paterson. By 1814, Tucker, Davis and Swan were travelling to Sydney to sell surplus wheat. In 1818, Macquarie visited Paterson's Plains and, presumably owing to the positive outcome of his emancipist experiment, allocated a further handful of 100 acre grants to emancipists at Wallis Plains (West Maitland).[18]

In 1819, Macquarie sought permission from British Secretary of State for the Colonies, Earl Bathurst, to close Newcastle convict station and instead transport offenders further north to Port Macquarie, clearing the way for a Hunter community focused on commercial development – such as winegrowing.

There were several factors at work in Macquarie's thinking. The Hunter's cedar was all but cut out as calculations of inexhaustible timber quantities had been overconfident. Crown rights to coal mining were increasingly being allocated to private investors. And Newcastle convicts knew the overland route back to Sydney via Windsor. Macquarie advised Bathurst that in addition to removing the temptation for convicts under

FOLLOWING PAGES:
Joseph Lycett, Lake Patterson, near Patterson's Plains, Hunter's River, New South Wales 1824. Plate 22 from Views in Australia or New South Wales and Van Diemen's Land, published by John Souter, London, 1824–25 hand-coloured aquatint and etching printed in dark brown ink 16.8 x 26.2 cm (image) 22.7 X 31.5 cm (plate) 27.0 X 36.8 cm (sheet). National Gallery of Victoria, Melbourne, Joe White Bequest, 2011 (2011.363).

new sentences of punishment to desert Newcastle for Sydney, he would like to see wealthy settlers (as distinct from emancipists or free settlers of humble means) established along Hunter's River, 'where they would have the combined advantages of a fertile soil of comparatively easy cultivation' and the ability to send their produce downriver to Newcastle, as a way station for coastal shipping to the principal market of Sydney.[19]

From 1821, in accordance with Macquarie's vision for the Hunter, large land grants of 2000 acres were made available to individual investors – prompting a flurry of requests from prospective landowners to the colonial offices in London and Sydney, on the basis of lines on a map and hopes of a fortune. Between 1821 and 1825, as many as 269 large grants in the Hunter received formal approval. A map of these allocations, drawn up in 1926, shows the filigree veins of the branches of the Hunter, Williams and Paterson rivers and of Wollombi Brook, crowded with the precise geometric lines and right angles of survey mapping where grants abutted each other along the full length of river frontages. Although other Hunter land would be apportioned after 1825, by that year the river frontages of the Hunter River had been comprehensively privatised – allowing settlers under British law to exclude Aboriginal people from this land if they chose with threats, fences and guns.

Back in Sydney, new immigrants who arrived from Britain to occupy their land grants in places like the Hunter crowded into the outer offices of the governor's official rooms, entreating gubernatorial favour by flattering his immediate staff. Men hoping for colonial advancement through paid government employment (like James Busby) joined other petitioners at Government House. Irrespective of their wealth or power, all new landholders were deciding how to invest in cultivation of their land at this cusp of a new era. (In response to a clamour of questions to this effect in the confines of his suite of offices, Macquarie's successor Governor Thomas Brisbane impatiently urged all new arrivals to 'Go to Hunter's River and make your fortune growing tobacco!', as a native plant resembling this 'royal herb' flourished there).[20] Many new settlers were committed to improving their land grants through farming and grazing, using convict labourers, and some picture had emerged of the sort of crops they should grow and sell. Survival crops for food supplies in the colony were established as of 1821, making experimentation in plantation-style luxury crops more viable.

Why plant wine grapes in the Hunter?

THE PRESENCE OF A CATALOGUED AND ACCESSIBLE COLLECTION OF *V. vinifera* at the Botanic Gardens in Sydney in 1827 (see Introduction) is a clue to the timing of vine plantings in the Hunter. Although the given names of grapevines would have baffled grape growers in established wine countries – as they were focused on wine trade taxonomies, rather than plant species names – there was some classification of vine type according to fruit characteristics and plant vigour. V. vinifera has many hundreds of cultivars (some of which are named in a different way in different wine countries). When preparing to plant cuttings to make a vineyard, the identity of a cultivar and whether it is suitable for wine cannot be determined simply by looking at the cutting, which is an anonymous length of two-year-old vine cane pruned from dormant vines after fruiting. These cuttings are obtained by pruning a parent vine with desirable characteristics – a method of propagation that replicates the parent plant. By contrast, propagating grapevines from seed leads to unpredictable results, as the DNA in seeds is programmed to vary from the parent plant.

Yet there was more to the planting of vines in the Hunter from 1828 than the availability of (seemingly) reliable stocks. The answer to the question 'why wine grapes in the Hunter?' was a combination of imperial ideologies and aspirations, and theories about plant cultivation that were not yet formalised into science. First, there was a concentration of early settlers in the Hunter who were influenced by currents of thought about winegrowing as a higher form of cultivation than some other crops; a symbol of respectable, civilised settlement. While they did plant survival crops like grains for flour, they also sought to make their mark with crops of greater value and prestige. Second, British knowledge about where to introduce European crops had arisen over several centuries of trial and error involving soil types and the need to match crops planted in colonies with the latitudes where these crops were grown in Europe. These ideas are outmoded now, but were cutting edge in their day.

A higher form of cultivation

WHEN COMPARED WITH OTHER TYPES OF HORTICULTURE, SUCH AS ANNUAL crops grown by ploughing soils and planting anew each year, wine grape vineyards have earned a reputation as a rarer, and therefore more exclusive, form of cultivation. Not all vineyards have this reputation, but the rarity and association with greater civility occurs in places where there are other vineyards and wines for comparison. Winegrowing requires substantial human intervention. The plants are perennial and the value of particular wines from specific places accumulates over time – sometimes over many centuries. For a start, whether cultivated for wine or eating fresh, or for drying into raisins or sultanas, grapevines take three years from first planting to bear a first crop, and vineyard owners must be able to afford a delay in income. Advice on winegrowing in colonial New South Wales held that after the initial cost outlay for a vineyard (land, vine stock, stakes for trellising tree-style, and labour – not including fencing), a landowner would recoup the investment after four grape harvests. This calculation depended of course on the vines being productive after three years, and then producing grapes that were suitable for wine or other purposes, and the product subsequently finding a market. The value of the vineyard land, and the grapes, depends on the availability of drinkers prepared to pay certain prices for wine.

Since antiquity, the transmission of wine-drinking culture has commonly preceded or accompanied the transplantation of wine grapes from wine countries to new wine worlds. When, more than two millennia ago, imperial Romans entering Gallic territories at Marseilles introduced *V. vinifera* to France, knowledge of which cultivars were most suitable for wine had been developing since the empire of the ancient Greeks, and later in the Middle East and eastern Europe, along with a culture of wine as elite and sacred. In Roman Gaul, Germania and elsewhere in western Europe, there were quite likely at first small, cosseted vine plantings – and trials, and errors – supported by those regional commanders or settlers who could afford to do so, just as a few wealthy settlers began the traditions of Hunter wine.

In the era in which British officials planned and executed the colonisation of Australia, the ruling classes of England and Scotland were prodigious wine drinkers. Wine lubricated their lust for colonial territories and for wealth from trade with these colonies, and wine symbolised their yearning for power.[21] Drawing on the symbolism of grapes and wine in the former great classical empires of Greece and Rome, British imperialists hoped that vineyards and winemaking could be achieved in the temperate climate of the colony of New South Wales.

Early Hunter winegrowers knew the names of many wines, and had opinions about the wines they preferred, as do drinkers today. There were those wines mentioned earlier in this book's Introduction: port, claret, sherry, Madeira, Canary and Constantia wines. Other wines traded by the British included Tokay from Hungary, and some Rhenish or German whites known as Hock, as an abbreviation of *Hochheimer*. At the turn of the 19th century, the middle and upper classes of the burgeoning British empire consumed wines according to their rank and income. Many British ships carried common wines instead of beer or ale as a preventative for scurvy. The more traditional British brewed beverages could be made onboard ship, but when travelling in certain regions the accessibility, and preservability, of wine proved greater than that of beer. In the British maritime world, trade wines had to be kept sound on long voyages through hot climates. This led, for example, to the characteristic caramelised sweetness of Madeiran wines – which resulted from heat and agitation in crossing and re-crossing the equator in barrels, in the holds of ships. Wines available for British consumption were the product of trade and diplomatic relations; and so, for example, Italian wines were not prominent, while British world ties with Portugal led to the ubiquity of port and Madeira.

Contemporary viticultural (grape growing) knowledge holds that grapevines for wine do not reach maturity until they are 10 years old, and once they are 45 years old their bearing potential is reduced to unprofitability (although the vines may still produce grapes of fascinating complexity). To maintain grape yields in a vineyard, vines must be replaced by the time they have reached 50 years of age. This shows the importance of wine places – whether those places are single vineyards or, more commonly, wine regions, such as the Hunter. Building a reputation for wine quality over time is not a factor of individual vines; it results from continual generations of production within a geographical space associated with wines. When new settlers planted vines in the Hunter from 1828, they could not know what the future held, but they did drink highly prized wines and they knew that these wines had some connection to long arcs of time.

A 'better sort' of settler in the Hunter

As the British colonial office intended, the availability of new 2000 acre land titles in the Hunter and elsewhere in New South Wales from 1821 attracted wealthy settlers who would intensify the pace of the production of colonial goods through farming and grazing. Compared with most of the emancipists of the colony, who had formed the larger population of landholders up until this point, the new wave of settlers gained their large and expensive grants on the condition of employing convicts and establishing colonial estates. These men and their families were of the educated, monied British middle classes who possessed political and moral authority, and therefore respectability, in their homeland; and this authority accompanied them to New South Wales. Some of these settlers aligned themselves with a later political movement of 'exclusives' in the colony, in opposition to the landholding, educated emancipists (former convicts and the offspring of convicts, whose social standing was tied to their parents), who sought to shape the colony's political landscape. A broader church of Hunter settler winegrowers did emerge later, but as the social conditions of New South Wales recast access to land and labour, wealthy emancipists rarely appear in the early phase of our history of the

Hunter wine community. Vineyard and winery owners in the region were chiefly wealthy settler capitalists – such as James Webber, a vocal member of the exclusives – or retired military officers.

In 1826, British-born James Atkinson, a New South Wales south coast settler who had spent four to five years in the colony, returned to London to a clamour of curiosity about progress in the antipodes. After being continually pressed with queries, Atkinson published *An Account of Agriculture and Grazing in New South Wales* (1826) to answer the questions he received. According to Atkinson's *Account*, 'the alluvial lands in New South Wales are not surpassed in fertility by any in the world' – principally 'along the banks of the Hawkesbury, Nepean and the various branches of Hunter's River' – with soils of 'vegetable mould more or less mixed with sand of many feet in depth'; although floods continued to be a risk for farmers in each of these districts.[22]

Atkinson advised that 'alluvial land upon Hunter's River is, generally speaking, in the hands of a better sort of people, by whom it will be managed with somewhat more intelligence and industry', in contrast with the emancipist and relatively poor 'dungaree settlers' of the Hawkesbury–Nepean who preceded the opening of land titles to the British middle class.[23] Atkinson's pronouncement on this 'better sort' of settler captures the tenor of capitalist entrepreneurship at Hunter's River. (Atkinson's remarks also reflect his social station – and his obliviousness to the fact that the natural endowments of portions of the Hawkesbury region were less conducive to European-style farming than the parts of the Hunter region made available to, and rapidly taken up by, wealthy and influential landowners. Karskens considers that Atkinson's criticisms were challenging the legitimacy of emancipist ownership of land compared with that of free settlers. She argues that Hawkesbury settlers were in fact productive and responsible for making the colony self-sufficient in grain.[24])

Upper Hunter settler Peter Cunningham – a naval surgeon – also believed his district of the Hunter Valley to be 'one of the most respectable' because of the concentration of British gentry and former officers.[25] Early settlement in the region became a rollcall of the New South Wales establishment: gentlemen by birth, university-trained clergy, medicos and lawyers, and those who had risen through the ranks of the military. These were the men in the colony appointed to the magistracy and the legislature, and employed as surveyors and in other paid positions as officials of empire. They were also merchants, shipping operators and industrial entrepreneurs. These 'better' settlers formed the Hunter's River Agricultural Club in 1827.[26]

In his *Account*, Atkinson commented obliquely on winegrowing. He stated that knowledge in service of all colonial cultivation remained pitifully deficient, yet a 'few intelligent gentlemen' had carried horticulture 'to a much greater degree of perfection than its agriculture', and others had recently invested in grapevines and olives 'with much success'.[27] Atkinson warned prospective New South Wales settlers that while Mediterranean cultivations such as wine grapes were desirable, they were expensive to establish. He also reassured his readers that the colonial future of 'new fruits and other productions' had an ideal champion in Charles Fraser, director of the Botanic Gardens – who liberally distributed plants to 'those persons likely to take care of them'.[28] Aspiration to winegrowing also required some notion of where exactly to plant vines, knowing that many risks of failure – and a long wait for fruit from the vines – attended the creation of a vineyard.

Suitable soils

SOILS THEORY IN THE BRITISH WORLD AT THE TURN OF THE 19TH CENTURY held that certain commercially grown plants preferred certain grades of soil fertility. Grapevines were believed to be best suited to sandy soils that were poor in nutrients and through which water flowed freely, preventing damp roots. This notion arose in part from an economic argument holding that loams which were too infertile for annual crops, such as wheat or corn, could be planted with grapevines. This would not only achieve a greater profit per acre, but would also increase land value – a sort of wasteland-into-wine idea.

The British men who created Australian maps were doing much more than plotting distances and ruling straight lines onto tanned animal skins and paper to formalise property ownership. They were kneeling on the ground, pushing aside the plants that grew there and inspecting the earth – seeking clues as to which profitable plants might replace the existing native vegetation that appeared to them to have no economic value.

As the colonial frontier of farming and grazing extended to Hunter's River, some doubts were emerging, as reported in the colonial press, about the efficacy of the 'rich soils for grain, poor soils for grapevines' concept, based on vineyards in and about Sydney. The dry, stony soils close to Sydney varied from Europe's sandy soils – in nutrient content for a start – and rainfall in New South Wales varied in timing, frequency and magnitude from rainfall patterns in the agricultural lands of Britain and Europe.[29] However, in the absence of deeper knowledge, the 'waste soils for wine grapes' theory persisted among some hopeful growers into the second half of the 19th century.

Likely latitudes

ANOTHER THEORY GOVERNING ENGLISH AMBITIONS FOR COLONIAL cultivation – that of climate parity – developed in English North America in the 17th century. Climate parity asserted that crops that thrived at certain latitudes north (above the equator) in Britain and Europe would flourish at the same latitudes north in the Americas – later, this 'parity' would apply to latitudes south in the antipodes. As it happens, the Portuguese wine-producing colony of Madeira is located at approximately latitude 33 degrees north, while Sydney is close to latitude 34 degrees south. This greatly bolstered confidence in the promotion of winegrowing in and near Sydney – despite the subtropical climate.[30]

It cannot have escaped the attention of colonial authorities in London or New South Wales colonists that the Cape colony in southern Africa is at latitude 33 degrees south. The fame of that region's wine (along with acquisition of its vine stocks) gave some encouragement to New South Wales wine aspirants. Yet it should be borne in mind that the colony of New South Wales was established not in a vacuum, but within the realpolitik of empire trade. In 1824, during the formation of the Australian Agricultural

Company, the company directors met with Colonial Secretary Bathurst to secure his support for their joint stock charter. The directors indicated their intention to cultivate wine grapes in New South Wales as a small part of the company's broad charter. Shortly after this meeting, however, Bathurst expressed the following concern: 'While appreciating the importance of the cultivation of fine wool and the olive, as part of the proposal for the Australian Agricultural Company, the establishment of vineyards on a large scale might interfere with the Cape of Good Hope's only viable export'.[31] The British had taken over possession of the Cape in 1806, and measured the colony's worth on the basis of its economy.

In 1826, the president of the Agricultural Society of New South Wales (and experimental winegrower), John Jamison, referred in his annual report to 'our parallel of latitude 34'.[32] The key matter here is that the Hunter Valley lies within the seemingly ideal latitude for colonial wine production and – irrespective of Bathurst's concerns – the Australian Agricultural Company, and other settlers, were determined to take advantage of this.

Later, as farmers gained experience of the weather patterns of agricultural districts in Australia, as the broader patterns of climate were better understood, and as the science of meteorology emerged, the blunt application of the concept of climate parity fell away. In the 1820s, though, the location of the Hunter Valley, and the wider region's natural endowments, were perceived as ideal for winegrowing.

Aboriginal Country and wine country

THE INSTIGATION OF HUNTER WINEGROWING OCCURRED AMID FIERCE frontier warfare. Aboriginal people raided colonial properties, burned crops and killed and wounded settlers; and settlers are recorded as raping and murdering Aboriginal people. In 1826, Governor Ralph Darling formed the first mounted police force in New South Wales, at Wallis Plains, from among the soldiers in the colony's British regiments. Early in 1827, the commanding officer at Wallis Plains, Lieutenant Nathaniel Lowe, authorised the shooting of an Aboriginal man named in settler records as Jacky Jacky, who was in custody at the Maitland lockup.

Governor Darling subsequently ordered Lowe's arrest and trial for the murder of Jacky Jacky, but a military jury declared Lowe not guilty. Historian John Connor argues that this verdict opened the way for mounted police to shoot Aboriginal insurgents on sight, if the police believed the locals had committed a felony – and Connor documents that between 1827 and 1834, the mounted police pursued operations against the Wiradjuri and Kamilaroi people of the Upper Hunter.[33]

During his travels in the Hunter, Dumaresq met a settler who kept 'the skull of a black fellow who was shot dead with a pistol ball in the act of making his escape from a party of police' as a warning to other Aboriginal locals to stay clear of the property. Dumaresq reported that colonial police sought a pair of Aboriginal men for insurgency, including 'Jerry', who is thought to have given his name to Jerry's Plains near Singleton, a district that today connects the Lower and Upper Hunter wine regions.[34]

In the Hunter, as elsewhere in Australia, the ancestral boundaries of Aboriginal people changed before and after colonisation, and are contested. It is for Aboriginal communities to define precise historical tribal borders in relation to contemporary vineyard sites, and so there are no claims here about *V. vinifera* being first planted in the land of a particular clan or band.[35] Aboriginal locals who were not killed or forced away from colonial estates are a part of early winegrowing stories until the 1840s, after which they become less visible within the Hunter winegrowing community – but this does not mean they disappeared from the region.

Aboriginal people did not cede their rights to their Country, yet under settler law they were accorded property rights in Australia only in recent decades. There has been at least one Aboriginal-owned vineyard – at Murrin Bridge Aboriginal Reserve in the New South Wales central west – but not in the Hunter.[36] In the first settler generation of winegrowing, from 1828 to 1846, Aboriginal people built relationships with settlers, and found ways to stay with the Country by sitting down on their estates, making them a constant presence in the Hunter, and on several winegrowing properties.

Chapter 2

A network of instigators: imagining Hunter winegrowing, 1828–35

IN THE EXTENSIVE GROUNDS OF TOCAL AGRICULTURAL COLLEGE, IN the Paterson district near Maitland, there are wide, shallow furrows covered with short grasses and low leafy plants, stretching across an area known as Line Paddock. From a distance, this field, on a gentle slope beside the steep upward curve of a modest ridge, appears similar to any other part of the former colonial estate of Tocal, established by James Webber in 1822. At close quarters, however, much of the earth of the paddock is revealed to be evenly rippled, like the swell of ocean waves, with spaces between the crests measuring about 4 metres and troughs of around 30 centimetres in depth. This is understood to be a site of Webber's vine plantings from the 1820s, using cuttings from the Sydney Botanic Gardens.[1] (Webber unlikely received sufficient vine cuttings at first to create the vineyard of 14 acres that can still be seen today. The vineyard may have been expanded over some years.)

Webber was a wealthy man, and more than 20 convict labourers were at first assigned to his employ as a condition of his receiving Tocal as a land grant. These convicts lived at Tocal, and they created the wave-like corrugations of the land while trenching the soil in readiness for laying out and planting vine cuttings. Trenching involves digging out the earth to a certain depth to loosen packed sediment; a contentious practice among colonial New South Wales winegrowers because of labour intensity and expense. Of course, the friability and fertility of soils also varied from estate to estate, and while some winegrowers found trenching to be essential to loosen the soil to boost vine root growth, others had no need of this practice, or preferred not to wear the great cost.

If vine cultivation had been less important to Webber, he would not have ordered the arduous process of shovelling trenches before vines were planted. The undulations in Line Paddock (thought to be named for the nearby railway line) are a symbol of Webber's desire to master winegrowing as a profoundly desirable (and picturesque) form of colonial cultivation and fruit processing. Grapevines are long gone from Line Paddock. The furrows remain because recent generations of college principals have recognised their significance as evidence of settler culture on the landscape.

Tocal was a key site in the first phase of settlers' picturing winegrowing in the Hunter, and Webber a key translator of turning these imagined hopes for vine cultivation into reality. Other first generation imaginers included John Pike and Samuel Wright in the Upper Hunter, William Kelman at Patrick's Plains, George Wyndham at Branxton and James King near Raymond Terrace. These settlers and others pictured vines on land where this crop had not been tried before. First generation winegrowers were fewer in number and more widely dispersed throughout the expanse of the Hunter region than is the case today. It must be said too that in these early decades there were many women involved in winegrowing by association: as the heads of the domestic world of their settler estates, as offspring of winegrowers, and as labourers. In the same way that convict workers trenched or purled the soil in preparation for planting the Tocal vineyard, Webber and other early winegrowers trialled ideas and practices that were to have ripple effects across succeeding generations.

The imaginers

IN 1832, NEW SOUTH WALES SUPERINTENDENT OF CONVICTS FREDERICK Hely (himself a vinegrower on the central coast near Sydney) recorded a list of Hunter settlers and their vine acreages. Hely made these notes in the flyleaf of his copy of *A Manual of Plain Directions* (1830),[2] by James

Busby. Like the furrows in Line Paddock, Hely's copy of the *Manual* is a key to understanding the earliest steps taken by Hunter winegrowers, the stages from thinking and talking about planting vines for wine, to indicating where a vineyard would be planted, and how large the vineyard would be. Unlike annual commercial plants such as corn or wheat, where some adjustment may be made year after year to the size and location of a crop, vineyards (like orchards), are a longer game. Vines must be protected and controlled until the first grape crop three years after planting. In the early colonial era it was a risky business not only to plant a vineyard, but also to expand vine acreage until there was evidence that a grape variety was suited to a specific location – and that the grape was the variety thought to have been planted.

According to Hely's list, there were about 15½ acres of vines in the Hunter in 1832. Webber's vineyard measured 3 acres. Other settler capitalists whose large properties hosted burgeoning vineyards were (radiating out around the compass from Tocal): George Townshend (2 acres at Trevallyn, Paterson); Alexander Warren (½ acre at Seaham, near Raymond Terrace); Henry Incledon Pilcher (1 acre at Telarah, near Maitland); Henry Dumaresq (1 acre at St Heliers, Lochinvar, west of Maitland) and Wyndham (2 acres at Dalwood on the Hunter River at Branxton); Kelman (1 acre at Kirkton, near Singleton);

Frederick Hely's list of early vinegrowers in the Hunter in his copy of Busby's Manual (1830). Dixon Library (DL) collection, State Library of NSW, Call no. 634.8/B copy 1.

Pike (1 acre at Pickering, near Denman); and Francis Little (1 acre at Invermien, near Scone). Fifty years ago, historian WP Driscoll drew on Hely's list to conclude that 'there is no indication of who actually planted the first vines in the Hunter Valley' – by which Driscoll meant one soul who might be attributed with solely pioneering Hunter winegrowing.[3] Hely's list in fact underscores that from the outset winegrowing encompassed a network of instigators; a community of characters, all of whom knew each other as a very small group of elite men whose families were at the helm of a collection of colonial estates.

When James Webber ordered vines from the Botanic Gardens in Sydney on behalf of the Hunter's River Agricultural Club in 1828 (as discussed in the Introduction), who else was in the club? A newspaper report indicates that, in addition to Webber, the club's members were James Phillips – whose property also fronted the Paterson and bordered Tocal to the north – and Timothy Nowlan, who held the Paterson-facing property to Webber's south. Neither of these settlers appear in later records of winegrowing, but other members of the club who do are Edward Gostwyck Cory, based further upriver at 'The Vineyard' (known too as 'Vineyard Cottage'), and Alexander Warren – an employee of the Australian Agricultural Company, with land at Seaham on the Williams River. Unfortunately, a newspaper report on the club does not give the names of other members.[4] Cory and Warren may not have cultivated vines from 1828, but they were among the men who constituted the generation of imaginers up to 1846. Other elites engaged in winegrowing were the Scott brothers, Andrew Lang, Charles Boydell and Alexander Park, whose stories are included in the following survey of early settlers.

James Phillips Webber, Tocal (Paterson)

JAMES PHILLIPS WEBBER EPITOMISED THE TYPE OF SETTLERS WHO considered themselves to be natural leaders in New South Wales. Although born in rural Wales, Webber's parents were of significant pedigree in imperial wealth and authority. His father hailed from an English family who profited from the colonisation of Ireland in the 17th century, and his mother's parents were wealthy Dutch-Americans who sided with British loyalists in the American War of Independence. Presumably Webber sought to head a new generation of his family in the antipodean colonies. Webber secured his initial Hunter grant, which he named Markham, in London in 1821, and immigrated to Australia in 1822 – placing him at the vanguard of new capital attracted to the colony.

On arrival in the Hunter, Webber renamed his property Tocal, an Aboriginal word meaning 'plenty' or 'bountiful'. (No record exists to confirm this, but Webber's concession to Aboriginal meaning in naming his property is an unexpected and intriguing start in the Hunter for a man with such a strong family history of imperial advancement in other English colonies.) Between 1822 and 1825, Webber increased his landholdings at Paterson to 3300 acres. According to historian Brian Walsh, by 1828 – when Tocal's labour force was at its peak – the estate was larger than average, employing 34 assigned convicts and two free men.[5]

As Walsh points out, in the 1820s much of the Paterson district comprised alluvial river and creek flats, dense with rainforest, and thick with vines and gigantic figs, cedar and gum trees. Tocal's hilly uplands were open woods, lightly timbered as a result of Aboriginal burning regimes that removed brush understorey. [6] Basalt soils suitable for native grasses are found among the eucalypt woodland portions of these uplands. Based on the presence of basalt soils in Line Paddock, historian Cameron Archer has deduced that at the time of settlement this parcel of land was kept grassed through fire farming, making it ideal for Webber's vineyard – in contrast with more thickly forested areas that required the removal of trees, stumps and rocks before crops were planted.[7]

We know that in 1834, Webber's vineyard contained Oporto and Gouais vines, as cuttings of this kind were sent that year from Tocal to Wyndham at Dalwood.[8] Both varieties could well have been from the Botanic Gardens' collection in Sydney. White Gouais Champagne was present, and bearing freely, in the Botanic Gardens in 1827; and Tinta, which also appears on the garden's list, is a grape variety grown in Portugal's Douro Valley for wine sold from Oporto.[9] Such exchanges of vine cuttings within the Hunter are scantily documented but create a reliable image that may be generalised across the social network of men such as Webber and Wyndham. In this way vineyards were expanded through cuttings shared between social peers, before the later sale of vine stocks within the region.

The Scott brothers of Glendon (Singleton) and Ash Island (Newcastle)

IN 1821, BROTHERS ROBERT, HELENUS AND (ALEXANDER) WALKER SCOTT arrived in Sydney from Bombay in India, where their father had served as a medical doctor, presumably with the East India Company. Scott senior died unexpectedly at the Cape during the immigration voyage, leaving the brothers sufficient capital to each obtain land grants of 2000 acres, a base from which they created a constellation of colonial properties for farming and grazing.[10] Robert and Helenus combined their grants in the estate of Glendon at Patrick's Plains, and resided there together. Walker established an estate at Ash Island in the Hunter River delta, later purchased Bengalla in the Upper Hunter, and ordered vine cuttings from William Macarthur.

Like most high-ranking gentlemen of this era, each of the Scott brothers were exceptionally mobile, frequently travelling to Sydney and elsewhere within the colony to visit peers, to assess the improvement of their assets, and generally to participate in the commotion of settler occupancy of New South Wales.

Edward Gostwyck Cory, Gostwyck (Paterson)

EDWARD CORY IMMIGRATED TO NEW SOUTH WALES IN 1823, WITH HIS wife Frances and his father John. He and his father each received a 2000 acre land grant fronting the Paterson River, which they named, respectively, Gostwyck and Cory Vale. Edward Cory rapidly developed his land holdings at Paterson, grazing sheep and breeding livestock. In the 1830s, he invested in a water mill for grain and a few years later built a steam-powered mill for Gostwyck. He subscribed to the publication of Busby's *Treatise*, but otherwise little is known of his winegrowing, or of his

Vineyard at front of homestead. Gostwyck, New South Wales, estate of E.G. Cory Esq, between 1834–51 / lithograph by George Rowe, Cheltenham, England, State Library of NSW, Call no. DG SV1B / 6.

father's enterprise. Edward Cory's landholdings for a time extended beyond the Hunter – although he resided in the Paterson district and reportedly sported the title 'King of Paterson'.[11] Cory was ultra-conservative in his politics, to an extent that set him apart even from contemporaries such as his close neighbour and friend James Webber.

A disturbing picture emerges from the colonial record on John Cory of Gostwyck. In 1827, Webber, as district magistrate, stationed two mounted police at Gostwyck to apprehend two Aboriginal men in relation to the non-fatal spearing of a convict, reportedly as payback for the killing of a dog belonging to the locals. A newspaper claimed that in a settler reprisal in connection with the spearing, 12 Aboriginal people were killed at Gostwyck by Cory's men; which Cory denied. As Brian Walsh argues, although the event remains 'a matter of conjecture ... the possibility of such devastating [settler] retaliation cannot be dismissed'.[12] Local historian Lesley Gent also draws attention to Cory's travails with escaped convict bushrangers, and the case of a convict assigned to Cory being hanged for striking his master with a shovel.[13] Although little is known about the alleged attacks on Cory by bushrangers, other Hunter stories suggest that bushrangers did not prey upon masters who were fair to their convict servants – which raises the distinct possibility that Cory did not treat his labourers well.

Andrew Lang, Dunmore (Maitland)

ANDREW LANG ARRIVED IN NEW SOUTH WALES WITH HIS PARENTS IN 1823. He immigrated some years after his older brother George, the original grantee of Dunmore, located between the Hunter and Paterson rivers. When George died in 1825, Andrew Lang assumed ownership of the property. By 1828, the estate supported a labour force of one female convict, 12 male convicts and two free labourers. Lang and his father constructed a horse-powered flour mill at Dunmore, and later invested in new windmill technology to more efficiently grind grain into flour.[14] These mills were constructed to process grain from several properties, and flour production must have been profitable to justify the update in equipment. Winegrowing at Dunmore remained less a source of income than an aspiration.

Andrew Lang's brother, John Dunmore Lang, a colonial clergyman, was active in several emigration schemes to boost the colonial population, including early agitation for his brother to gain government subsidies to import experienced labourers for the Dunmore vineyard. In the 1830s, John Dunmore Lang's ambition to access skilled workers for his brother's winegrowing enterprise was surpassed only by the efforts of the children of John and Elizabeth Macarthur – John, Edward and William, in particular – to obtain permission and funds from colonial authorities for non-British immigrants from winegrowing countries in Europe. The Macarthur brothers did not at any stage own land privately in the Hunter, although their brother-in-law James Bowman did. (If Bowman experimented with winegrowing, then little is known about this.) Nevertheless, the wide-ranging colonial business interests of the Macarthur brothers were a source of continual comparison and competition for John Dunmore Lang, his family, and others.

In a revised edition of John Dunmore Lang's guide for immigrants to New South Wales (first published in 1834), the Dunmore vineyard is described as 8 acres in extent, planned and formed by George Schmid – 'a highly intelligent Wirtemberger' – into four compartments with paths intersecting at right angles, sloping to a centre point with an arbour of vines called a *lusthaus* (pleasure house). In southern Germany, Lang had apparently seen families taking their meals in these agreeable, shady constructions.[15] This description is embroidered by Lang or another writer, as it varies greatly from his earlier text. It is to be hoped that the words are Lang's, for they are revealing of the Hunter winegrowing community.

Tellingly, in the revised version of his guide the reverend gentleman Lang strongly disapproved of the effect of Hunter wine on lower class intemperance, as labourers reportedly bought wine by the bucketful directly from winegrowers.[16] Reverend Lang remarks that labourers were ideal customers for winegrowers in the absence of middle class drinkers of any number in the colony. In addition, the sale of any product from an estate to its workers as in-kind payment had long been legal. In the early years of the colony, Lang actively supported winegrowing as an antidote to the inebriety of colonial labourers – which did in fact peak in the 1830s.[17]

In later published versions of Lang's immigrants' guide, therefore, some accounting had to be made for the fact that colonial wine became a source of drunkenness as much as any other form of alcohol. It is worth mentioning this later reference to buckets of wine, even though an unwary reader of Lang might assume he wrote it in the 1830s. It offers some sense of the colonial reputation that grew around wine in the Hunter, from the reality of finding drinkers during the long game of determining how to make wines for more discerning customers.

William Ogilvie, Merton (Denman)

A RETIRED NAVAL COMMANDER, WILLIAM OGILVIE ARRIVED IN SYDNEY IN 1825 accompanied by his wife Mary and son Edward. His many assigned convicts were sent to their Upper Hunter land grant, Merton. Visiting Merton in 1826, Cunningham found that the property's 6000 acres (expanded from the original 2000 acre grant), between the western bank of the Hunter River and the eastern bank of the Goulburn River, boasted 'alluvial flats and lightly timbered forest land ... bounded by a moderately high ridge'. Riding further into the property, 'A plain of fifty acres of rich land (*without a tree upon it*)' – Cunningham's emphasis – 'is situated in the middle of the grant, overlooked by a beautiful swelling hill, equally clear, of the finest sort of garden mould, and covered with luxuriant grasses'.[18] Cunningham appreciated the beauty of this part of the valley, as:

> Contrary to what is generally found in other parts of the country, the ridges upon the upper part of the Hunter's River are almost uniformly flattened at the top, forming little miniature hills and valleys covered with fine soil of moderate depth, and abounding in grass, which makes it the great resort of the kangaroos and cattle in the winter season.

Cunningham's own grant was also located in the Upper Hunter, and convict labourers assigned to him were managed in his absence by Ogilvie.[19] According to William and Mary's daughter Ellen Ogilvie, the first vines at Merton were planted by Mary, and William later hired a German man (with the surname Luther – Ellen did not record his first name) as caretaker of the vines and as winemaker.[20]

John Pike, Pickering (Denman)

JOHN PIKE'S PROPERTY, PICKERING, LAY ON THE EASTERN BANK OF THE Hunter, across the river from Merton. During the 1820s, Pike extended his original 2000 acre grant to 6000 acres with holdings at Wybong. Pike served in the New South Wales military garrison under Governor Macquarie until 1821, after which he served in the British garrison at Ceylon (Sri Lanka), before returning to New South Wales in 1825 with his wife and daughter – and in the company of fellow officers Ogilvie and Cunningham. Pike's friendship with Ogilvie and Cunningham, and new connections with Wyndham and others, placed him at the centre of a shared vision for vines and colonial wine underwritten by success in grazing and breeding sheep and in raising grain crops. With extensive land in the Upper Hunter, Pike appears to have continued his pastoral pursuits for longer than contemporaries further south. Pike remained in the colony until 1859, then sold his property and travelled to Europe.

Francis Little, Invermien (Scone)

IN 1823, SCOTTISH SURGEON FRANCIS LITTLE APPLIED FOR A NEW SOUTH Wales land grant on the recommendation of his uncle, colonial surgeon William Bell Carlyle, and immigrated in 1825 to occupy the 2000 acres of Invermien. Little does not appear to have continued with grape growing after his initial planting of 1 acre, or at least his efforts are not visible in family correspondence or other letters and publications.[21]

Henry Dumaresq, St Helier's (Lochinvar)

BROTHERS HENRY AND WILLIAM DUMARESQ WERE JERSEY-BORN MILITARY men who immigrated to New South Wales in 1825. Henry's brother-in-law

Governor Darling (married to a sister of the Dumaresq brothers) had offered Henry a position in the colonial administration, and William (a year younger) accompanied him to the antipodes.[22] Henry's distinguished military career included serving as an officer at the battle of Waterloo against Napoleon. Consequently, Henry received high-level government appointments in the colony before establishing an estate on his land grant near Lochinvar, Hunter's River, in 1827. William took on roles such as surveyor while also developing an Upper Hunter land grant, which he named St Aubin's. The brothers both extended their property holdings to close to 6000 acres apiece. Henry and William were the subject of conjecture by Darling's colonial enemies about nepotism, during Darling's governorship from December 1824 to October 1831. Henry also attracted approbation for the quality of his estate management at St Helier's – maintaining order among his convict servants by separating unmarried convicts by sex, and by being a kind master.

Henry Dumaresq's letters do not, unfortunately, detail his grape plantings. They do however reveal an observant and flamboyant correspondent who longed for his mother to join him and his wife and children in New South Wales.

In 1827, Dumaresq wrote to his mother about St Helier's, explaining that the intended homestead site lay 'about a Quarter of a Mile to the Right Hand' of a branch of the Hunter River, 'with a Magnificent alluvial Flat intervening in front', surrounded by swamp oaks. Between his house and stables, 'is a delve or little Glen' where he planned an orchard and garden. 'There now', he asked his mother, 'how do you think you should like it – and what do you say to assisting me to beautify it?'. Although the description does not contain any allusion to vines, Dumaresq's honeyed phrases of a hopeful future offer a view of settler intentions amid the bounty of the alluvial districts. 'My Dairy, Wool House, Shearing Shed and Farm Buildings are to be about a Quarter of a Mile from the Cottage, up the Glen or Vale', he continued, and:

> now to do all this – the Stone is Quarried on the Spot – the Lime is found in abundance – the Timber is cut close to the Saw-Pit – the

River and Brook furnish Fish and Wild Fowls in myriads – and you [collect] your Coals from the Banks of the former – without more ado than to shovel them up – as the action of the Water has left them exposed.²³

Henry, like his brother William, shared enthusiasm for describing the landscape as bountiful in its present state – and yet ripe for colonial improvement through transformation into a British-style farm.

Dumaresq complained to a friend in 1831 about the behaviour of convict servants (which is curious in the light of his reputation for being a good master), adding that: 'My faithful Blacks, who have lived with me 10 years make us very independent of the Larcenous Tribe of Servants here [the convicts]; and we are in this respect much better off than our neighbours'.²⁴ Unfortunately more is not known about the Aboriginal people Dumaresq refers to. Still, the remark intimates that Dumaresq's success in managing convicts arose from his lesser reliance on them compared with some other fellow settlers.

Speculatively, Aboriginal people may have been involved in the formation of Dumaresq's vineyard, reported by Hely in 1832 to be 1 acre in extent.

If Dumaresq made wine at St Helier's, there were few opportunities. In 1834, he was appointed commissioner of the Australian Agricultural Company and he died in 1838, reportedly from the lingering effects of a bullet wound received as he delivered despatches at Waterloo.²⁵

George Townshend, Trevallyn (Paterson)

GEORGE TOWNSHEND IMMIGRATED TO NEW SOUTH WALES IN 1825, AGED in his late twenties, to take possession of one of the 2000 acre land grants being dispensed in the Hunter district. For this younger son of a large family of Welsh gentry, the colonies offered riches unobtainable in Britain. Townshend shared his passage to Sydney with his Welsh neighbour, 17-year-old Charles Boydell, who was also seeking new economic opportunities. Townshend and Boydell travelled to New South Wales in the company of Stephen Barker (hired by Townshend to manage his colonial estate for three years), settler capitalist Alexander Park, a group of Germans recruited by the Australian Agricultural Company as shepherds, and sheep purchased by the company for their Hunter Valley grant.[26]

Townshend is a curious inclusion among early Hunter winegrowers, as it seems he did not intend to make wine at his Paterson Valley land grant; however, he was growing grapevines in 1832, which earned him a place on Hely's list.

Townshend became friends with James Webber, which added a further strand to the regional web of vine cultivators. Townshend also delivered vine cuttings, presumably from his vineyard, to George Wyndham at Dalwood. (Wyndham states that he received cuttings from Townshend, but not whether they were from Trevallyn or elsewhere.)

In August 1833, Wyndham recorded that Townshend brought him: 'White Muscat, Black Damascus (large fleshy), White Gouais, White Muscatel (good eating), Shepherd's white, Captain Anley's Red Muscatel, Large Purple, Corinth, Wantage, Black Frontignac, Madina, Tinto'.[27] Of especial interest in this collection is the variety 'Shepherd's white', as this is the earliest reference to Semillon in the Hunter (known for many years in the colony as Shepherd's Riesling or Hunter River Riesling). It indicates that Townshend likely obtained his stocks directly from Shepherd, as did many others in this period. Thomas Shepherd arrived in Sydney in 1825 from England via New Zealand, where he had been associated with a failed joint stock enterprise (a New Zealand version of the Australian Agricultural Company). Shepherd's expertise as a nurseryman earned him great respect, and a good deal of affection, within the community of settler agriculturalists in New South Wales.

The other varieties on the list represent the typical colonial assemblage of stocks from many different sources. Several are likely to have been varieties from the Botanic Gardens from 1828, or perhaps during the dispersals of stocks from the gardens organised by James Busby to accompany the sale of his *Manual* – in 1830. The brief mention of Captain Anley points to the existence of other grape plantings beyond those listed by Hely in 1832, although perhaps Anley cultivated at a smaller scale than some settlers.

George Townshend's estate. Trevallyn, 6 October 1836, Conrad Martens, Scenes in Sydney & New South Wales, 1836–63, State Library of NSW, Call no. PXC 296, frame 29.

During the economic Depression in New South Wales in 1840, Townshend was declared bankrupt and lost Trevallyn and other properties he had accumulated in the Paterson and Gresford district.

Charles Boydell, Camyr Allyn (Gresford)

CHARLES BOYDELL WAS THE FOURTH SON IN A FAMILY OF 15 OFFSPRING – this may have inspired him to join Townshend in immigrating to become a landowner in New South Wales in 1825, rather than settle for landless opportunities in Wales. Boydell seems to have started with fewer financial resources than his friend Townshend. Boydell acquired Camyr Allyn, on the Allyn River, as a 2000 acre land grant in 1826. But the younger man developed his holding more slowly than did Townshend (an impulsive character) – adding crops and fences over several years.

In January 1833, Boydell visited the Upper Hunter and reported that Pike's vineyard contained 'ripe' and 'abundant', 'Black and White Cluster, Sweetwater and Burgundy', along with some other fruits. He also commented that Townshend's Sweetwater grapes bore a tremendous crop that year, despite Townshend choosing not to 'train' or trellis the vines, but rather to leave them to 'grow upon the ground like gooseberry Bushes'.[28] These details appear in the pages of Boydell's diary, alongside a list of words he learned from an Aboriginal man he referred to as Jacky. Jacky may have belonged to the Aboriginal group whom Boydell and his family harboured at Camyr Allyn. Boydell, like Dumaresq, wrote

Diary of visit to fellow winegrowers and Aboriginal lexicon on facing pages. Charles Boydell, Journal 1830–35, State Library of NSW, Call No. A 2014, pp. 51–52.

that his enterprise got along well because of Aboriginal people who, at Camyr Allyn, assisted with a tobacco harvest. Perhaps these locals also planted and pruned vines, and picked grapes.

In 1836, Charles's younger brother William arrived in Paterson, becoming the owner of Caergwrle, another property associated with early colonial winegrowing. Charles Boydell went on to take his place among the colonial establishment as a district magistrate, and Camyr Allyn remained in family hands until 1911.

Samuel Wright, Bengalla (Muswellbrook)

DURING THE NAPOLEONIC WARS, BRITISH TROOPS JOINED FORCES WITH the Spanish military to prevent a French conquest of the Iberian Peninsula. The Peninsular Wars ended in French defeat in 1814. Samuel Wright, a veteran of these wars – as a captain of the Third Foot (Scots) Guards – arrived in New South Wales in 1822 with his regiment. From 1823 to 1825 Wright served as commandant at a Tasmanian convict station, and in 1826 he was commandant at Port Macquarie convict station, on the mid-north coast of New South Wales. In 1827, Wright received a grant of land near Ogilvie's Merton in the Upper Hunter (the precise acreage is not clear), and applied to sell his military commission in order to retire. While awaiting the sale of his commission he served as police magistrate at Newcastle.[29]

In the early 1830s, Wright and fellow military retiree John Pike provided vine cuttings from their own vineyards to Wyndham. The vines in one such exchange were Red Portugal (this could be any number of grape varieties from the Portuguese Douro Valley or elsewhere – including Tempranillo, as it is known in Spain), Green Malaga (perhaps a white grape used to make Muscatel), Constantia (from the Cape), Black Cluster, Black Damascus, Alexandra Frontignac (a Muscat grape) and Black Sweetwater (most likely an eating grape, although early colonists experimented with making wine from all varieties).[30] This list contains some grape types possibly received from the Botanic Gardens, alongside others with names that also connect them with the Douro Valley, Madeira and the Cape. It is not immaterial that serving in Iberia may have allowed Wright access to Spanish and Portuguese winelands; regardless, as Englishmen, both he and Pike would have been familiar with wines from these regions. If Wright continued to grow grapes at Bengalla past the 1830s, or to make wine, no records remain.

The sharing of vines by Wright, Pike, Wyndham, Townshend, Webber and others shows how these settlers shared a vision to make the Hunter a wine region, alongside other commercial enterprises.

Henry Incledon Pilcher, Telarah (Maitland)

IN 1830, HENRY PILCHER, A LONDON BARRISTER, ARRIVED IN SYDNEY with the intention of becoming a large-scale wine producer, and in possession of a manuscript of his translations of French methods of winegrowing. The *Sydney Gazette* reported that Pilcher was 'highly respectable'; however, the editor had regretfully declined to print Pilcher's manuscript of 'practical as well as scientific knowledge' about winegrowing (written as a result of 'tours upon the continent, and a careful observation of the various modes in which vineyards and the manufacture of wine are conducted there'). The newspaper had that very day agreed to print James Busby's *Manual* ('far more concise and practical' than Busby's 1825 *Treatise*) – and advised Pilcher to 'postpone publication to a more convenient season'.[31] How disappointing for Pilcher! Despite the rejection of his publication, Pilcher proceeded to grow vines, and make wine, and occupied a central role in the winegrowing community for some time.

> In 1843, a battle between Aboriginal locals occurred close to Henry Pilcher's property.

In contrast with the settler capitalists on their large grants, Pilcher and his family lived on a 60 acre town block called Telarah, and continued to acquire further town properties. In 1843, a battle between Aboriginal locals occurred close to Pilcher's Telarah property. The dispute between the 'Maitland tribe' and 'Port Stephens tribe' resolved in favour of the 'Maitland tribe', who were carrying muskets.[32] Such an event may have been a form of traditional dispute resolution or a type of competition, rather than warfare. It is not surprising that muskets were incorporated into this ritual battle as there was evidence of trustful sharing of firearms between settlers and locals in Newcastle in the 1820s, and in the Lower Hunter in the 1830s.[33] The locals' battle signals continued Aboriginal presence near Maitland and the incorporation of settler technologies into local lives.

Telarah adjoined the Hungerford family's property of Lochdon. Eliza Pilcher, daughter of Henry and Eliza, married Reverend Septimus Hungerford – a branch of the Hungerford family tree that later connects with Pokolbin's Tyrrell family, who enter our history during the second settler generation of winegrowing. Pilcher belonged to Hungerford's congregation of evangelical Anglicans, who appear to have embraced wine cultivation and moderate wine consumption as an acceptable element of their colonial civility.[34]

Alexander Warren, Seaham (Seaham)

A TRAINED AGRICULTURALIST, ALEXANDER WARREN ARRIVED IN NEW South Wales before 1824, although the exact year is not clear, and he may have immigrated expressly as an employee of the Australian Agricultural Company at Port Stephens. Warren received a 1000 acre land grant at Seaham on the Williams River, which is notable for being half the usual acreage allocated to settlers at this time. As James Busby – a friend of Warren – wrote to William Kelman, the shortfall in acreage was owing to increasing pressure for grants in the Hunter, a few short years after the opening of applications for land. Busby offers no theory for the apparent discrimination against Warren.[35] Perhaps Warren's lower social standing played a role. In the 1820s, even as a new wave of middle class officers and gentlemen arrived in New South Wales, recasting the social and political milieu, conditions in the colony were bringing changes to notions of class, power and influence based on birth and professions. Some years later, Warren would take his place in the close circle of Hunter winegrowers, with no hint that he might be of lesser status.

Lewinsbrook, 1853, rows of grape vines in the foreground, coloured grey. William Leigh, Coloured sketches – South Australia, New South Wales, New Zealand, Cape of Good Hope, Image no. 42. Lewinsbrook. The residence of A. Park Esq. MLC. State Library of NSW, Call no. PXA 1987.

Alexander Park, Lewinsbrook (Gresford)

AS A PROFESSIONAL MAN, SCOTTISH-BORN MEDICO ALEXANDER PARK HAD no trouble acquiring a sizeable land grant in the Lower Hunter – over 2000 acres near Gresford, within the Paterson district. Park arrived in Sydney in 1826, aged 21 years, and soon afterwards established his property, Lewinsbrook. He may have been a member of the Hunter's River Agricultural Club when Webber placed the 1828 order for vines from the Botanic Gardens. Park's brother is understood to have been a merchant in the Madeiran wine trade, which might explain how Park came to employ two Madeiran Portuguese workers at Lewinsbrook as early as 1827.[36] Park

Another view of Lewinsbrook. William Leigh, Image 43. At Lewinsbrook looking over the Valley. Oct 28/53. State Library of NSW, Call no. PXA 1987.

may have employed the workers at Madeira on a layover during his immigration voyage to Australia, as the Sydney emancipist and surgeon William Redfern had taken this step earlier in the 1820s.

In his *Manual*, Busby credited Park, along with Townshend and the Australian Agricultural Company, with importing Verdelho to New South Wales, and 'several of the other varieties which are cultivated at Madeira'.[37] John Macarthur – a leader of the Australian Agricultural Company venture – imported vine stock from Madeira on his return to New South Wales from long exile in 1819, having travelled through France to spend a year in Switzerland with his sons James and William – to complete their education, and to gain knowledge of winegrowing.[38]

In the 1875 edition of Lang's guide for immigrants, Lewinsbrook vineyard is described as being 35 acres in extent. (Lang mentioned too that the famed British expeditioner in Africa, Mungo Park, was Alexander Park's uncle.)[39]

The scale of all types of cultivation at Lewinsbrook – and Park's status as a professional gentleman – made him a natural supporter of campaigns for the importation of labourers from British India to replace convict labourers, once transportation of felons to New South Wales ceased in 1840. From 1853 to 1868, Park served as a member of the New South Wales Legislative Council for three separate terms, in the seat originally held by Richard Windeyer (who appears in chapter 3). It is possible Park advocated for colonial winegrowing during the introduction, in 1862, of a licence fee of £1 per annum for wine shops, as opposed to a licence fee of £30 per annum for public houses.

Aboriginal locals at Gresford are known to have sought out Park when in need of settler medical attention, as he treated Aboriginal people more sympathetically than did some other medicos.[40]

George Wyndham, Dalwood (Branxton)

GEORGE WYNDHAM IMMIGRATED FROM ENGLAND IN 1827 WITH HIS WIFE Margaret. He purchased the 2080 acre property known as George Home, fronting the Hunter River at Branxton, which they renamed Dalwood.

William Leigh, Sketches in New South Wales, 1853, Image no. 28. Germans Hut Lewinsbrook, State Library of NSW, Call no. PXA 1988.

In 1830, Wyndham planted vines that he had received from James Busby's distribution of stocks from the Botanic Gardens, but these died – this may have been due to Wyndham's inexperience or because of a problem with the plant stock. The following year, Wyndham purchased orchard stocks from Thomas Shepherd, and vine stocks from Gregory Blaxland's Brush Farm in Sydney, and planted these with more success. By June 1833, he recorded his vineyard as 4 acres in extent. In 1835 – in great anticipation – Wyndham made wine in two repurposed hogshead barrels (a hogshead holds 40 gallons of liquid). Although this first vintage proved disappointing, he could not be deterred from imagining that drinkable wine could be produced from his vineyard.[41]

Wyndham numbered among the few settler capitalists who could afford the construction of a winery. Gavi Duncan, a descendant of Darkinjung man William Bird, remembers family stories telling that when 'George Wyndham built a winery, our mob helped to build that winery', as part of the labour they performed at Dalwood. 'Even though Wyndham operated under European law', according to Duncan, the settler 'considered the land still belonged to Aboriginal people and treated them fairly with wages. My middle name is Wyndham, so is my father's middle name'.[42] Like his wealthiest peers, Wyndham owned several colonial properties, and his sons and employees later trialled winegrowing in other places in addition to Dalwood, especially Bukkulla at Inverell in the New England region of northern New South Wales.

Wyndham was garrulous by temperament and, freed from much of the actual physical toil of land ownership (the tasks performed instead by Aboriginal people, convicts and free labourers), he spent much time in sociable conversation in parlours and paddocks. Those whom Wyndham visited, or who came to Dalwood, included Webber, the Scott brothers (at Glendon and Ash Island), Dumaresq, Pike and Wright, Cory, Townshend, Ogilvie, and two others we have yet to introduce – William Caswell and John Glennie – giving a concrete sense of a social network in operation that benefited the progress of settler industry.[43]

A principal division among the winegrowing imaginers in the Hunter lay between those for whom winegrowing was a small and novel part of their enterprise, underwritten by additional sources of income such as pastoralism in other districts or breeding livestock (especially horses), and those settlers whose livelihoods depended more acutely on income from grapes and wine. While Wyndham could be counted among the former, William Kelman belonged to the latter category. Kelman had a smaller capital base, and greater hands-on involvement in the running of his farm.

William Kelman, Kirkton (Singleton), and the Busby family

WHEN IN 1823 SCOTSMAN WILLIAM KELMAN SAILED FROM ENGLAND TO Van Diemen's Land to occupy a land grant, he shared the voyage with a family of fellow Scots: minerals surveyor and engineer John Busby, his wife Sarah Busby and several of their adult children. John Busby had travelled to take up a government post he had actively sought.[44] During the voyage, Kelman became friends with Catherine Busby, daughter of John and Sarah, and with her brother James. As discussed in the Introduction, James Busby was a young agriculturalist trained in Scotland or England, with experience of living in Bordeaux, who had been inspired by an 1821

British parliamentary report on colonial industries to write an instruction manual on French vine cultivation and winemaking for readers in New South Wales.

The Kelman and Busby voyage laid over at the Cape in 1824. In January that year Kelman wrote his mother a letter in which he described a 'delightful' tour of the Constantia wine estate, undertaken with Catherine and James. 'When we arrived at Constantia', Kelman began, 'we were very handsomely treated by the Owner of the House who seems to be a Healthy man [,] he took us through his Gardens and Vineyard & showed us all the rarities of the place, where there is [sic] all the fruits of Africa growing spontaneously'. Following on from this pleasant ramble, the estate owner 'treated us with a glass of the different kinds of Wine which are white & red Constantia & Frontignac [,] the finest wines in the World & which we were sure were not adulterated'.[45] This whole experience played well for Kelman; the health (or respectability and sobriety) of the estate owner, the famously picturesque estate, and the thrill of drinking fine – and unadulterated – wines at their source.

Kelman completed his agreement to transform his Van Diemen's Land grant into a working farm, and with the assistance of James Busby he received a new land grant (although where is not clear from the records), in New South Wales, in 1828. This situation allowed Catherine Busby to marry Kelman and remain in the same colony as her parents and siblings. However, Kelman did not choose to live on his own grant in the Hunter, which lacked a fresh water supply. Instead, he and Catherine lived on her father's estate at Kirkton, which offered many natural advantages – including proximity to water. In 1831, Kelman resigned his original land in favour of leasing property nearer to Kirkton, and continued to reside at his father-in-law's property.[46]

In 1829, two of Catherine's other brothers, Alexander (Ellick) and William, established their Cassilis property. Only Ellick lived at Cassilis, William remained in Sydney, acting as an agent for his siblings – a role held by men from other notable families in settler history. The Busby family shared their resources, including convict servants as well as grazing land. The Busby property at the border of the Hunter region gave its name to the Cassilis Gate, the flow-through point for transport from the Hunter to the western slopes and plains beyond the Great Dividing Range. Cassilis Gate is a less formidable route from the coastal plains to the Australian interior than the perilous colonial roadway across the Blue Mountains near Sydney, but as Sydney became the sorting house for most colonial trade, Cassilis Gate did not provide access to as much settler traffic as its natural features afforded.[47]

In 1829, Catherine wrote to her brother George Busby, a doctor at Bathurst, that she and her husband had built a 'very good house' at Kirkton. Once finished, it would be 'very respectable'. It had a fenced-in garden of 2 acres, in which William Kelman 'works a great deal'; being 'very fond of gardening'. There were more than 300 sheep at Kirkton, 12 acres of wheat (with plans to plant maize), and assigned convicts were fencing a new 60 acre paddock.[48]

> The vineyard acreage that Hely recorded was most likely planted out with vines from James Busby's Botanic Gardens cuttings.

The vineyard acreage that Hely recorded for Kelman in 1832 was most likely planted out with vines from James Busby's 1830 distribution of cuttings from the Botanic Gardens.

In 1833, Ellick Busby expressed a determination to plant his share of newly imported vine stocks from James Busby at Cassilis, but awaited the return of a convict labourer to assist.[49] In July of 1833, William Busby advised his sister Catherine that:

> Tomorrow I intend putting onboard the Wm the fourth [the coastal packet] a case of vine cuttings among which is a set of the French stocks imported by James. The numbers refers to the Vineyard Collection in James' last publication [Journal of a Tour]. You will find cuttings in some instances where I could not find an Original but often you will find a Duplicate. I have kept a Duplicate of the whole set here. Have sent a set to James and sending a set to Ellick and Sempill [at Belltrees, Scone] but yours is the best lot next to that we have Kept. I would not presume to tell such a Gardener as your Good Man is, how to plant them, but I will tell you how I am doing

mine. I have cut the numbers on stakes which I will fix besides the vine whose number it bears. And this I can always tell by reference to the book what vine blights, or bears luxuriantly. If K. plants them in his Vineyard let him do it not as a Vineyard but as an Experimental garden whence he can plant the cuttings of the vines which suit the climate best, into his vineyard. Tell him that I consider them a very valuable present.[50]

This letter establishes that Kirkton received vines that included stocks collected by James Busby in Europe – although the list William Busby sent to Kirkton, 'among which' were vines from James Busby's French collection, were in fact:

blk Cluster, blk Oporto or Portugal, 1 Stone Grape from Mr Campbell, Large red grape with only 1 seed, Very Large blk Grape like an Egg, Varieties imported by Capt Wilson, best large White grape, black Hamburgh, Miller's burgundy, White Muscatel eating grape, White Gouais, Sherry, Name unknown and Large White grape Mr Campbell.[51]

Rather than making the nature of Kelman's vinegrowing experiments clearer, this list clouds the issue. When William planned to send a 'set of the French stocks', it seems he meant one grape variety among the many listed here, and there is no clear indication of which one. A Red Hermitage grape later valued by Kelman may (very speculatively) have been the French stock referred to. If so, it would have required a great deal of patience over many years to produce sufficient vine cuttings – through the pruning of established grapevines – to make a large enough vineyard for commercial wine production.

Also in 1833, in July, a Sydney settler asked Kelman's advice about procedures with vine cuttings. 'Will you please have the kindness to inform me whether I may put my vine cuttings into the beds this year and transplant them next season', asked R Peters, 'as I have the ground where they can remain so as to thrive … gravel in a good situation but at present wheat is growing on the ground'. Peters explained that 'I can't find anything about transplanting in Mr Busby's Manual so pray give me your opinion on the subject and if I put them out in beds how far each cutting should be from the other – how near do you put them for a trellis?'.[52] There are two things to be said about this letter from Peters. First, that publications on viticulture were not yet adequate for the purpose of advancing an inexperienced vinegrower to success. Also, that Kelman gained some authority on vine cultivation that has been lost to history because he did not publish vinegrowing or winemaking instructions as did James Busby, William Macarthur and some others.

In spring of 1833, a letter arrived at Kirkton from William Busby to William Kelman, advising that he would post by coastal packet some goods including six bottles of (presumably imported) wine. Busby wrote too of their common interests in sheep breeding and wheat growing, of sharing convict labourers between properties for seasonal harvest, and of trouble with convicts.

William Busby wrote to William Kelman in mid-November 1833. 'I am glad to hear that your Vineyard is succeeding so well. 'What do you require to make Wine? I wish you could manage it this year.' Following his signature, Busby gave further thought to the matter, writing that, 'I have sent a cask instead of a box it may be useful for making Wine'.[53]

Clearly then, in 1833, Kelman had not yet made wine.

Busby and Webber, departure and legacy

JAMES WEBBER AND JAMES BUSBY ARE CLOSELY ASSOCIATED WITH EARLY Hunter winegrowing – Webber through his role in spurring early vine plantings and Busby as a result of his promotion of vine cultivation colony-wide, and later mythmaking. Yet by the mid-1830s neither of these men remained in New South Wales. Webber left the colony as a network of dedicated winegrowers began to emerge in the Hunter by which time Busby had taken up a new appointment as official British Resident in New Zealand. Webber's departure appears to have resulted from his inability to obtain a foothold in the colonial ruling class. He belonged to the so-called exclusives, who opposed former convicts' rights to hold property or official postings in the colony, and he campaigned vehemently against emancipists in the press under the penname OQP. Webber considered himself to be a natural leader, but did not reckon on a social structure where emancipists were educated rhetoricians, landowners and newspaper proprietors who could take control of political debate.[54]

Whereas many others in the New South Wales ruling class were upwardly mobile men with origins in the lower ranks, Webber's parents were from powerful families within the British empire; and in immigrating to the colonies their son had intended to become a great plantation owner. This is evident from the catalogue of sale for Webber's Tocal library, of which 250 of the 350 volumes concerned agriculture or gardening or, in the case of *The Distiller's Guide*, instructions about turning surplus or damaged grape harvests to profit by making spirit alcohol.[55] It is a fascinating exercise to try to understand what Webber might have learned from these many tomes and applied in designing the vineyard that his convict workers so deeply inscribed onto Line Paddock.

Notably, the Tocal library contained volumes one and two of French botanist Henri-Louis Duhamel du Monceau's *Elements of Agriculture* (translated into English by Philip Miller in 1764). In volume one, Duhamel cautions against the over-manuring of grapevines as a substitute for regular 'tillage'. 'Nothing is so striking', warned Duhamel, 'as the different quality of Wine produced from an undunged Vine, and that made from the grape of one much dunged'.[56] The decision to trench for vines may have arisen from reading Busby's *Treatise*, or – quite likely – from conversations with other vinegrowing enthusiasts. As for the dung, the lack of livestock grazing in the Hunter likely took care of that – supplies of dung were not widely available in Webber's time in the colony. But Duhamel's recommendation for regular tillage may have convinced Webber of the need for broad vine spacing in the trenches of his vineyard.

In 1834, Sydney merchants Caleb Wilson and his son Felix purchased Tocal estate, as landlords. An English tailor, Wilson had immigrated to New South Wales with his wife and two-year-old Felix in 1804, only to lose his wife on the voyage. Over subsequent decades father and son acquired considerable wealth through trade. After Webber's departure from Tocal, long-term tenant Charles Reynolds continued grape cultivation there. The homestead constructed by Webber was destroyed by fire in 1835, and in 1841 Felix Wilson built a two-storey mansion at Tocal, although he still did not reside at the estate.[57]

The grand Tocal homestead remains open to the public today at the site of Tocal Agricultural College – established through a bequest from the last private owner of the property. Among the features Wilson incorporated into the design of the house is an ornate fireplace resplendent with a pair of neoclassical carvings: grape bunches on long-stemmed plinths, framing the hearth. Apart from reflecting vine cultivation at Tocal, the ornamental grapes recall a British mercantile fondness for such representations. Grapes and grapevines as decorative forms were deployed from the late 18th century in Britain to evoke ancient Greek and Roman bounty, prosperity and civility; to dignify emerging British power.[58] As John Dunmore Lang mentioned casually of an encounter with smallhold settlers in the Hunter in the late 1820s, '*the bush*' – his emphasis – meant 'uncultivated country'.[59] This uncultivated state meant not only meant the literal absence of European crops and gardens, but also a lack of civility, or manners, which the Wilsons were attempting to correct with classical symbols ornamenting the Tocal hearth.

Wilson served as a director of the Bank of New South Wales from 1843 to 1850, placing him among powerful members of the colonial establishment.

Webber's – and even Busby's – vision for winegrowing relied on transplanting expectations from British and European worlds, but by 1835 it remained for other settlers such as Wyndham and his social network, and Kelman, to maintain the energies to realise this vision.

Chapter 3

Experiments and expectations: willing Hunter wine into being, 1836–46

JAMES KING ARRIVED IN NEW SOUTH WALES FROM SCOTLAND IN 1827 suffused with a spirit of scientific entrepreneurialism, and in search of a colonial fortune. King's origins in trade and manufacture were at variance with those of settler capitalists from the English gentry, such as James Webber and George Wyndham, and of William Kelman the farmer. Whereas Kelman at Kirkton excelled in managing vines and making wine, King observed and classified – exhibiting an interest in geology, plant biology and the burgeoning of wine chemistry in Europe. At the same time he directed his mind's eye to customs duties and regulations governing wine business, and the further horizons of export from the Hunter to Sydney, with a view to later exporting to London. More than most other winegrowers, King thought in terms of reaching for and contributing to the development of scientific knowledge, and of forging trade pathways – even before there was much wine to trade.

King named his Hunter land grant Irrawang, which is understood to be an Aboriginal word for 'water view' – although this could be ironic, as King petitioned unsuccessfully for many years to gain frontage on the Williams River. Today, the location of his estate (divided by the Pacific Motorway just north of Raymond Terrace) away from other areas of intensive agriculture on the river flats near Maitland – as well as the property's sandy soils, hosting eucalypt forest – suggest that Irrawang at no time suited intensive agriculture.

What King lacked in richness of real estate he made up for in determination. He created one of the few early Hunter manufactories – a pottery that produced wares for serving and storing food and drinks – and his wine earned an international reputation through exhibition competitions, despite King never achieving profit from his wine business. King's social network among Hunter winegrowers comprised neighbours in the Williams River district and nearby Port Stephens. His energetic embrace of a collective approach made him a pillar of the wider community of vine cultivators and winemakers who were dispersed throughout the districts of the region in the first settler generation of winegrowing. King's associations beyond the Hunter were not as closely linked to political and military influence as those of some other colonists, but still extended into colonial halls of power. As they did with Wyndham, other settlers orbited around him in the strong bonds of social kinship that consoled settlers distant from their close ties with family and society back in Britain.

> King's energetic embrace of a collective approach made him a pillar of the wider community of vine cultivators and winemakers.

In late winter of 1832 and 1833, King's convict labour force laid out and planted Irrawang's 'experimental' vineyard of the 'most approved [grape] varieties', in soils of 'puddingstone and porphyry' – 'trenched, broken and turned over with the spade, to not less than 30 inches deep'. Puddingstone and porphyry are rocks made up of varied sized stones and minerals. In puddingstone the larger stones are dark, and may be crystalline, and in porphyry there are lighter coloured crystals. It is emblematic of King's keen attention to every element of the natural resources of his estate that he recorded his vineyard soils according to existing classifications.

Within a few years, King's vines 'grew vigorously' and fruited 'abundantly'.[1]

An etching by John Carmichael depicts Irrawang in about 1838 (the precise year is unknown) as an industrious world of crops, kilns and other outbuildings.[2] At the centre of Carmichael's image are convicts busy about the estate. In the background, the vineyard is a series of parallel rows across the slope of a hill. In the foreground are large urns labelled 'grapes', with bunches strewn around them, while more grapes on vines are entwined in a giant gum tree at the far left. Under the tree is grouped an Aboriginal family of man, woman and child, and a dog, gazing into the settler world of the estate. Carmichael's composition is aspirational, as the clean lines of the etching conceal the rudimentary nature of the enterprise. The vines in the eucalypt and the scattered urns evoke the frequent metaphors of vines and wine in the Old and New Testament and in classical literature. Carmichael's picture encapsulates the settlers' conviction of the rectitude and dignitas of colonial progress – especially through the introduction of grapes and wine.

In 1827, gentleman traveller Dumaresq had found settlers at Williams River sleeping with firearms close to hand, but Carmichael's etching indicates a

Two years after the first vintage at Irrawang. 'Irrawang vineyard and pottery, east Australia' [picture]/ J. Carmichael [?] c1838. Rex Nan Kivell Collection; NK1477, National Library of Australia.

new tenor to the district a decade later – the sort of progress in development that settlers such as King considered to be their calling. While Aboriginal people in Carmichael's image were certainly inhabitants of the wider terrain of Irrawang, King's property may not have been a specific site offering shelter and bounty to traditional owners, as there were other places where this could be more readily found, closer to water. The inference is that these Aboriginal locals at Irrawang were at the margins of settlement, drawn to a beneficial 'civilising' presence of settler industry; whereas, surely, for locals the development of Irrawang estate was strange, and hardly welcome.

In contrast with the gentry of the Lower Hunter, such as Webber, Wyndham and others mentioned in chapter 2, settlers at Williams River were often of middling rank and, like King, late 1820s immigrants to the colony. A social network developed between the Williams River district settlers and middle-ranked employees of the Australian Agricultural Company at Tahlee, who also owned farms. One link between the group of winegrowers gathered around King and those who associated with George Wyndham, as discussed in the previous chapter, was William Caswell, probably because of his former employment in the military.

William Caswell, Tanilba (Port Stephens)

FORMER NAVAL LIEUTENANT WILLIAM CASWELL IMMIGRATED TO NEW South Wales with his wife Susan and their children in 1828 to take possession of several land grants, including Tanilba – named for the Aboriginal word meaning 'white flowers'. Susan's brother, Robert Hoddle, had immigrated some years earlier, and is the better known member of this family for his role as a colonial surveyor. Caswell's daughter Emily married Andrew Lang at Dunmore, and Caswell and William Burnett (who immigrated for the Australian Agricultural Company) were colonial allies; if not prior to their colonial careers, certainly once they were both established in the Hunter. After Caswell died on a voyage back to England with Andrew and Emily Lang in 1859, Susan auctioned the family's possessions, including winemaking equipment, and returned to her homeland.[3]

Edward Parry and William Burnett, Australian Agricultural Company, Tahlee (Port Stephens)

EDWARD PARRY AND WILLIAM BURNETT IMMIGRATED TO NEW SOUTH Wales in 1829: Parry as Australian Agricultural Company manager and Burnett as the company's superintendent of agriculture. (Parry therefore visited Wyndham; Burnett was friends with King.)

When their voyage laid over at the Cape colony, Parry reported to the company's agent in London that Burnett 'will go to Constantia to view wines' – for business, rather than pleasure.[4] The initial delivery of vine cuttings to Tahlee is not recorded, but the two varieties cultivated there were known as Black Hamburgh and Black Cluster, or the claret grape (possibly Cabernet Sauvignon).

Early contact between Aboriginal people and settlers near Tahlee. Port Stevens [Stephens] N.S.W. looking East, Views N.S. Wales / [collection of eleven watercolour drawings by Augustus Earle], State Library of NSW, Call no. PXD 265.

First vintage

IN 1832, PARRY RECORDED A DESCRIPTION OF THE AUSTRALIAN Agricultural Company's vintage, the earliest known to have occurred in the wider Hunter region. According to Parry, on 3 February 1832 – a fine day with south-east to north-easterly winds and 'passing clouds' – he and eight others were engaged for 'some hours making wine'.[5] A fermenting vat fashioned from a 40 gallon 'harness cask' was washed out with boiling water and then saltwater, as advised by Busby (presumably in the *Manual*). The Black Hamburgh grapes were 'ripe and in beautiful order'. Starting at eight o'clock in the morning, convict workers picked and weighed the fruit, and pressed 250 pounds of grapes with stems. Another 40 pounds of grapes were retained for eating and 'some of the finest left on the trees' (perhaps to attempt the prized method of late-harvest winemaking). From 11 am until 1.20 pm, workers alternated, 'stamping or trotting on a board

put over the grapes'. The pressing produced 17½ gallons of wine, to which Parry added '2½ gallons of good moist sugar'. (Busby's *Manual* recommended adding sugar to young and under-ripe grapes.)[6]

And then began the anxious wait for fermentation.

On the second day after pressing, a light 'scum' (Busby also uses this word) rose to the top of the fermenting vat, and by the third day (Parry is careful to emphasise that this is 48 hours after harvest) 'a hissing sound was perceptible, and a few bubbles appeared'. Cellar temperature measured 70°F and the wine 72°F, and a wooden cover was placed loosely over the vat. On Monday, 6 February, the cellar temperature had risen by a degree; at first the wine remained stable, until the fermenting vat finally came alive with 'some continuous blisters as well as larger bubbles', and a hissing noise, by which time the taste of the wine began to change to 'somewhat vinous, very sweet'. By Tuesday, the wine temperature had risen to 75°F, although the atmospheric temperature had not changed. The wine tasted 'less sweet but still *very* much so' (Parry's emphasis). A day later the ferment still hissed. The 'head' had not risen but the wine 'tasted less sweet and much more vinous'. The sweet taste (from unfermented sugars) remained on Thursday. On Friday, the wine was drawn off into a cask, and over the next ten days, Parry oversaw bottling of the wine.

It may have rained during vintage. Parry left the Black Cluster on the vines: 'We could not pick them on account of the weather', he explained. But when he elected to harvest on Tuesday, 21 February – a 'moist sultry day, wind southerly' – the grapes 'were too far gone, I fear, to do any good'.

On 19 October, Parry and others tasted the wine, which they 'pronounced to be excellent' and 'very different indeed'. The wines were clear and Parry decided not to fine or clarify them. That is, to add eggwhite or another gelatinous substance to remove lees – and proceeded instead to bottling.

In total, 98 bottles were filled with the separate varietals and labelled according to the order in which the wines were drawn from the storage casks. The first drawn received the highest quality Australian Agricultural Company seal, the second wines drawn were labelled with Parry's own seal, and the third with a 'wafer stamp'. The Black Hamburgh was sealed with black wax and the Black Cluster with red.[7]

Tahlee, Port Stephens, 23 April 1841.
Album of Pencil Sketches, c1828–60 by
Conrad Martens. State Library of NSW,
Call no. PXC 299.

In 1833, the vintage at Tahlee began on 4 February.

There is some ambiguity in Parry's report about grape varieties compared with the previous year. In 1833, Parry harvested only the Black Cluster, as this grape cultivar 'produced much the best wine last year', and he may have been advised that Black Hamburgh is for eating fresh, not for making wine. At nine o'clock in the morning (no weather report, unfortunately) on the day of vintage, with the Black Cluster 'quite ripe, very large and juicy, and with very few bad ones', the grapes were gathered and stalks removed before pressing. Without stalks the grapes weighed 380 pounds. Work halted at sunset, and the following day a further 111 pounds of grapes were 'picked and pressed', occupying workers until 3 pm.

The pressing produced 40 gallons of fermented wine, and Parry added 1½ pounds of brown sugar to this wine in a 60 gallon fermenting cask – a barrel with the top cut out. Parry reports that 'fermentation began briskly', and then the foaming head fell a little on Wednesday before rising again with increased vigour.[8] A tasting at 8 am 'seemed to me still to be sweet, but by 11 am we all agreed that it had lost much of its sweetness, and had become decidedly vinous to the taste, and the head had fallen considerably'. Parry attributed the faster fermentation compared with the 1832 vintage to 'the larger quantity and the superior sweetness of the juice'. Again, fermentation temperatures varied from 70°F to 73°F.

Parry does not indicate how many workers were engaged in the 1833 vintage. From his account, we can see however that six or seven men removed stalks and two other workers pressed the grapes by foot. Parry thought that in future more hands ought to be allocated to picking and that two presses in operation could ensure the harvest occurred in a single day, to prevent inefficiencies.

The 1833 vintage was bottled on 3 October and sealed with red wax using the stamp of Henry Darch, Parry's secretary. A further eight bottles that were damaged were sealed with black wax.

> In January 1834, 30 bottles of the highest quality Black Hamburgh and a dozen bottles of the Black Cluster were despatched to London.

In January 1834, 30 bottles of the highest quality Black Hamburgh and a dozen bottles of the Black Cluster (ten with the Australian Agricultural Company's seal, two with Parry's seal) from the 1832 vintage were despatched to London. Parry reported to London that the company's vines were flourishing and that he was considering planting 500 more (about half an acre's worth), probably from prunings of existing vines. It is noteworthy that Parry recorded that his purpose in winemaking was 'not a source of profit but interest and a substitute for spirits'.[9] This reflects anxieties among colonial leaders and the clergy about alcohol overconsumption, because drunken workers diminished the efficiency of colonial progress and reflected poorly on the reputation of the colony within the wider empire. It is also possible that Parry addresses the concerns of Colonial Secretary Bathurst, expressed in 1824 (see chapter 1), that the Cape colony's wine industry might be threatened by wine production by the Australian Agricultural Company. It cannot be ruled out either that Parry was trying to avoid taxes on the wine when it arrived in London, as such matters greatly exercised wine producers, even from this time. There was a very real possibility of colonial enterprise being stymied by rules that were made to suit other circumstances – such as taxes on alcohol importations to collect government revenue.

In Parry's account, we can see that he viewed the experience of winemaking as tremendously novel and as a source of discovery because of the elemental nature of the know-how and equipment.

On visiting the terraced vineyard and new wine house at the Regentville estate of John Jamison (see chapter 1), Parry found that the property commanded a very pleasing view – but concluded that this estate's wines were yet to be perfected. 'Winemaking must be a matter of speculation and experiment at first', philosophised Parry, clearly enjoying his role in this speculative process.[10]

Henry Carmichael, Porphyry Point (Seaham)

SCOTTISH-BORN HENRY CARMICHAEL – A FRIEND OF JAMES KING, AND deeply interested in the philosophy as well as practicality of wine – immigrated in 1830 at the invitation of John Dunmore Lang to teach classics at Lang's short-lived Australian College in Sydney, modelled on the Belfast College of elementary and secondary education.[11] Carmichael soon fell out with Lang – it is not clear why this occurred, although both of these colonial characters could be sanctimonious. Carmichael later secured employment as an assistant surveyor in the Hunter.[12] Grapevines were planted at Carmichael's Porphyry Point property in 1838 and again in 1841 with advice, if not direct involvement, from King.

Carmichael emerges from the records of his speeches and correspondence as a talented yet impulsive rhetorician and advocate for winegrowing in preference to pastoralism and other land uses. His advocacy of winegrowing was likely tied to his knowledge of the literature of the classical Greeks and Romans, but had concrete connections to his documented reading habits. During the sea passage from London to Sydney, Carmichael had read aloud Adam Smith's *Inquiry into the Wealth of Nations* (1776) to lower class fellow travellers. *Wealth of Nations* is an Enlightenment creed prescribing free trade and industrial specialisation as the pathway for British national wealth and greatness, within the wider community of European empires that included the declining powers of Spain and Portugal. It is no coincidence, however, amid the politics of a preference for French wine among certain British intellectuals, that Smith also favoured free trade of French wine to Britain – at a time when trade policies on British wine imports discriminated against French products in favour of those of Portugal, charging lower tariffs for port and Madeira than for claret and burgundy. Smith had served as a private tutor with a family in France in the 1760s and developed an interest in French wine, which he shared with fellow Scottish philosopher David Hume. Smith's views on the civilising effects of winegrowing and wine drinking were surely integral to Carmichael's determination to succeed as a winegrower, although the difficulty of making profits from this enterprise later subdued his outspoken convictions.[13]

William Charles Wentworth, Windermere and Luskintyre (Singleton)

IN 1836, WILLIAM CHARLES WENTWORTH PURCHASED WINDERMERE AND Luskintyre on the Hunter River and became one of the region's best-known absentee landlords. Wentworth is a figure of great stature in Australian political history as a mid-19th century constitutionalist, parliamentarian and newspaper proprietor. Wentworth is thought to have been conceived before the voyage to Sydney in 1790 of his convict mother Catherine Crowley and his father, D'Arcy Wentworth, who travelled to New South Wales to be employed as a surgeon. This combination of respectable versus shameful origins made Wentworth something of an enigma at a time when birth became the marker of social station in the colonies, although there seems to have been less stigma attached to Wentworth's parentage in the more cosmopolitan salons of London. Wentworth grew up in New South Wales and then studied law in England. A year spent in Europe, including France, resulted in a friendship with the Macarthur family, although his convict origins prevented his marriage to a daughter of Elizabeth and John Macarthur. The Macarthurs were among the exclusives, wealthy settlers free of any convict background, who like James Webber sought to formalise class privilege as the basis of political authority in New South Wales in opposition to any perceived taint of convictism – something that Wentworth vigorously opposed.[14] For these reasons Wentworth is a different sort of imaginer again than others of this era associated with winegrowing.

It makes sense that Windermere and Luskintyre possessed vineyards, among other assets, prior to Wentworth's purchase, because of their location in the Hunter. Moreover, Wentworth's social connections extended to several of the most prominent New South Wales winegrowers. It is unclear whether Wentworth's sojourn in Europe converted him to winegrowing, although it is possible that this was so – he certainly did have a reputation for possessing a fine wine cellar.

Wentworth did not participate in person in the sharing of vines and knowledge among Hunter wine peers, but he did later support vinegrowing interests in the colonial legislature, and conceivably contributed to the reputation of Hunter wine in Sydney's high society.

Richard and Maria Windeyer, Tomago (Tomago)

RICHARD WINDEYER'S PARENTS IMMIGRATED TO NEW SOUTH WALES IN the 1820s, leaving Richard to complete his legal training in London. In 1835, Richard, his wife Maria and their infant son William followed Richard's parents to the colony, where he rapidly gained respect for his skills as a barrister, and earned a reputation for judicious attention to the rights of Aboriginal people. Richard Windeyer also became prominent in the colonial legislature.

In 1838, Richard purchased Tomago estate, and further extended his colonial holdings to become a major landowner in other districts. This portfolio of properties and Tomago's stately home and cultivations came at great expense, however, and when Richard died unexpectedly in 1847, aged only 41 years, Tomago was heavily mortgaged. Yet, in an era when women rarely owned property, Maria succeeded in purchasing Tomago from the mortgagee, with the financial support of her father and some former associates of her husband. Maria's two sisters and their children also lived at Tomago, which Maria managed frugally for many years. She maintained a German vineyard employee and other labourers, and supported her son's education at The King's School, Parramatta. William Windeyer became the first graduate of the University of Sydney, followed his father into the legal profession, and had a distinguished career as a parliamentarian. Some of William's reforms were focused on women's equity in property ownership, and his wife Mary Windeyer was one of the most prominent colonial suffragists of the late 19th century.[15]

Archibald Windeyer, Kinross (Raymond Terrace)

RICHARD WINDEYER'S UNCLE ARCHIBALD – THE BROTHER OF RICHARD'S father – immigrated to New South Wales with his wife Elizabeth and their eight children in 1838, and joined the educated colonial elite. In 1840, he purchased Kinross (perhaps at a mortgagee sale during that year's economic collapse in the colony), and his life followed the pattern of many other wealthy and influential settlers with property in the Hunter's River district and links into the wider pastoral lands and families of the colony.[16] Windeyer's Kinross estate located him within James King's close geographical sphere of winegrowing influence, as well as being about 8 miles from Tomago.

A reordering of the propertied class

IN 1839, A SYDNEY NEWSPAPER REPORTED ON HUNTER'S RIVER winegrowing by King, Caswell, Wyndham, John Pike, William Dunn (of Wollombi – about whom little is known) and William Ogilvie. According to this report, Lieutenant Caswell's Tanilba boasted 6 or 7 acres of vines, 3 acres of which were six years old, and 2 acres four years old. Caswell's 1839 vintage produced six hogsheads. King made ten hogsheads, 'and would have made twenty hogsheads but for the hot winds six weeks ago, which cut off half his vintage'. Kelman sent a supply of grapes to a Sydney merchant that were 'larger than any we ever saw or heard of, either in this colony or at home'.[17] Of the grape producers we have discussed, perhaps only Kelman made any income from grapes.

In 1840, most colonial wealth came from wool exports to Britain – but that year an international financial bubble burst and the trading price of wool crashed. This fall saw a sharp rise in insolvencies. The Bank of Australia failed, leaving settler investors, such as the Scott brothers, almost penniless.

This first colonial Depression arose from a series of factors. For a start, all arable land and native grassland had been allocated to settler investors in the Hunter Valley in the 1820s, but pastoralism could no longer continue in the places where it had begun now that the edible grasses had been entirely eaten out by sheep and cattle. (Nor were native grasses renewed by the Aboriginal fire farming practices that had created them in the first place, as settlers had disrupted these traditions.) Pasture crops had not yet been planted in the colony to feed livestock, and quick measures were required in order not to lose grazing animals as future income. Wyndham, for one, moved his livestock and his family northward to their other properties in the New England district where pastures were not yet depleted.

> An estimated 50 per cent of the colony's large investors felt the impact of the Depression, and many of these were based in the Hunter.

Additionally, after industrial booms in Britain and the United States, commodity prices fell, slowing the inflow of capital from those nations to the Australian colonies. And disastrous inflation in land prices, resulting from overheated investment, left landholders with mortgages higher than tenant rents would cover.[18] An estimated 50 per cent of the colony's large investors felt the impact of the Depression, and many of these were based in the Hunter. George Townshend was among the settlers bankrupted. Conversely, the failure of some settlers provided opportunities for new colonists to obtain valuable land. Among these newcomers was Henry Lindeman, who purchased part of Townshend's insolvent estate.

Henry Lindeman, Cawarra (Gresford)

HENRY LINDEMAN SERVED AS A SHIP'S SURGEON WITH THE BRITISH EAST India Company on the India to China route. In 1840, Lindeman and his new wife Eliza immigrated to New South Wales, sharing their passage with Charles Reynolds, who followed Webber as the proprietor of Tocal. Once Lindeman received permission to practise in the colony as a medical doctor, he and Eliza lived first in the Paterson district with Thomas Patch, another East India Company man, and Patch's family. Lindeman purchased Cawarra in 1842 from among the portfolio of Townshend's former properties. Cawarra is an Aboriginal word meaning either 'flowing water' or 'meeting place', although it remains to be settled who gave the property this name.[19]

Montague Parnell, Poole Farm (West Maitland)

MONTAGUE PARNELL WAS A MEDICAL DOCTOR WHO HELD A PROMINENT position in the burgeoning Hunter wine community in the second half of the 19th century. And yet he has been largely forgotten, because there were no second generation wine business purveyors in his family as there were among the sons of Henry and Eliza Lindeman. Born in Exeter, England, and trained as a surgeon in Paris, Parnell arrived in Sydney as a young man in 1842, and began a practice in Maitland. Parnell's life in France inspired his interest in winemaking (this is a matter of public record, whereas for some other settlers it remains speculation), and he received many awards from intercolonial and international exhibitions.[20]

Edwin Hickey, Osterley (Raymond Terrace)

IT IS NOT CLEAR WHEN EDWIN HICKEY ARRIVED IN NEW SOUTH WALES, but he purchased Osterley – located between Raymond Terrace and Hinton – at a mortgagee sale in 1842, and proceeded to pursue entrepreneurial opportunities. Hickey is credited with instigating the short-lived Hunter River New Steam Navigation Company, which connected Newcastle with Morpeth and Maitland by steamboat. Steam-powered vessels operated between Sydney, Newcastle and Port Stephens, and up the Hunter River, from 1831. Later, the colonial government invested in rail and road infrastructure, which from the mid-century gradually replaced river transport. Hickey also engaged in other profit-making ventures, including the production of olive oil and brickmaking.[21] He is the only member of the Hunter River Vineyard Association (see chapter 4) to have also belonging to the shorter lived New South Wales Vineyard Association (which could more accurately have been named for Sydney and the Cumberland Plain). An organisation controlled by William Macarthur, it was quickly subsumed into the broader Agricultural Society of New South Wales from which it had arisen, rather than existing as a separate entity.

Leichhardt visits Glendon, Kirkton and Dalwood

YOUNG PRUSSIAN NATURALIST LUDWIG LEICHHARDT ARRIVED IN NEW South Wales in 1842 with the intention of observing and documenting native plants and animals, minerals and rock forms, as an exercise in pure research – for knowledge's sake. This approach was distinct from that of settlers such as James King, whose concern with new scientific thinking was intended to assist his colonial enterprises. Although German-born, Leichhardt studied in England and benefited from the financial support of an English patron. Leichhardt is well known within explorer histories of Australia, as some six years after his arrival in Sydney, while on a celebrated scientific expedition in Queensland's Darling Downs, he and his party disappeared – leaving generations of schoolchildren with an enduring impression of the continent's sparsely populated interior landscape as bleak, harsh and unforgiving. In contrast with the image of Leichhardt as an ill-fated scientific adventurer, it is intriguing to consider that in the summer of late 1842 and early 1843 he spent many months in the bountiful coastal hinterland of the Hunter. During his time in the region – although his diaries and letters are occupied principally with recording the native environment – Leichhardt expressed interest in viticulture and winemaking, leaving wondrous descriptions of Glendon, Dalwood and Kirkton.

In December 1842, following a stay of some weeks at the home of Walker Scott at Ash Island, Leichhardt relocated to Glendon, home of Robert and Helenus Scott, and from there rambled to many other properties, appearing (he felt sure) as a peculiar figure to settlers: 'a stranger slung around with collecting canisters and boxes'.[22] Probably because of his idiosyncratic attire, and his earnest inquiries, Leichhardt enjoyed an enthusiastic greeting from George Wyndham, who was very fond of receiving visitors.

Visiting Dalwood in mid-December 1842, Leichhardt evoked the heat of the season. Wyndham's vineyard lay 'close to the bank of the Hunter on a sandy, mild but fertile elevation … surrounded by lemon trees', bearing fruit that was shrivelled summer-dry. Tellingly – given we might mistakenly consider that historic vineyards resembled the vineyards of today – Leichhardt described Wyndham's vines as planted 6 feet apart and in rows 6 feet wide. The vines appeared to be 'very sturdy … real trees more than six feet in height, as in Italian vineyards. The sticks are thick, square, rough', and many were up to ten years old. Because of a shortage of labour to tend the vines, Wyndham confessed that for two years he had allowed cattle to feed on them, and Leichhardt saw that the stock 'have almost grazed it bare'. Still, the wine house – 'a stone building serving as a cellar' –

In the midpoint of this sketch, beside the house is what appears to be a vineyard. 'The Hunter at Glendon', 3 May 1841, Conrad Martens, [Album of] Pencil sketches, watercolours, etc by C. Martens, et al, c1823–63, State Library of NSW, Call no. PXC 284, frame 6.

contained wine from earlier vintages, and Leichhardt found that some of this wine mixed with water 'really is a very pleasant drink'.²³

During his extended stay with the Scott family at Glendon, it was William Kelman, however, whom Leichhardt sought out with particular eagerness.

Arriving at Kirkton on the allotted day of his visit there, Leichhardt passed along an entrance track lined with aloe vera plants and large cacti, which gave the 'vineyard and house quite a tropical character'. Kelman, in contrast with the striking ornamentation of his driveway, was evidently less flamboyant than Wyndham and the Scott brothers, and Leichhardt regarded his host as the better studied horticulturalist. According to Leichhardt, Kelman's 'large' vineyard was planted about 40 feet 'above a gully which, after some bends, falls into the Hunter' at a stretch of the stream that remained flowing year-round. The vineyard soils were red and yellow sand mixed with a little clay and humus: 'loose, easy to cultivate'. The vine varieties were 'red Alexandria, red and white Hermitage, Cornuta, Chazelas and many other kinds'. Kelman ploughed but did not trench his land before planting vines. The vineyard at Kirkton had initially been laid out in the same style as Wyndham's – 6 feet apart in 6 foot wide rows and trained 6 feet high – but Kelman later revised the distance between the posts to 3 feet, and the width of the rows to 3 feet. Kirkton vines were staked to a height of 15 to 18 inches – the style of planting used in many vineyards into the early 20th century in the Hunter.

As for the dressing of the vines: rather than tying down the fruit-bearing vine canes to other stakes placed beside the planting supports (as some growers did), Kelman allowed the canes to hang from the main vine wood, 'to obtain good table grapes and these require plenty of shade'. Kelman pruned his vines to sit low, and while some growers pruned their vines twice a year, Kelman pruned once a year – in autumn – to maintain a leaf canopy against summer sunshine. (Leichhardt ventured that the Italian style of tree-trellising remained desirable in Australia's heat.)

Kelman's vineyard was surrounded by 'a beautiful cactus hedge' (as with the driveway, this was probably prickly pear – a common method of early colonial fencing), and a copse of Cape mulberry trees shielded the vines against drying northerly, north-westerly and westerly winds. The 1843 grape harvest produced 'beautiful' fruit, but yields were low because Kelman had removed the older long canes – and the new crop bore fruit from new shoots on old vine wood.[24] Reading between the lines, this means that Kelman had previously allowed the vines to take on the appearance of small bushes, with many fruiting canes. However, it seems he changed to limiting the number of fruiting vine canes, so that while each individual plant grew fewer bunches of grapes, the bunches were of better quality for winemaking.

In readiness for making wine, Kelman's 'fermenting-tub stood on a verandah close with boards, his wine-casks lay in a 4 foot deep vault-like hollow, over which a timber-hut had been erected, surrounded on three sides by a closed-in verandah'. This tub was fashioned from a very large wine barrel, with the base removed, and stood vertically, resting on another barrel. Kirkton wine-storage barrels were mainly sherry casks.

> Of all his grape varieties, Kelman made wine only from White Hermitage (Rousanne or Marsanne) and Red Hermitage (probably Shiraz).

Inside, the cellar felt cool compared with the summer heat outside, but lacking a thermometer Leichhardt could not provide a precise difference in temperature. Kelman described the process of fermentation of his crushed grape juice, or must, as 'very wild and violent', and generally 24 hours in duration.[25] Of all his grape varieties, Kelman made wine only from White Hermitage (Rousanne or Marsanne) and Red Hermitage (probably Shiraz).

Leichhardt approved highly of the white wine (which he described as greenish hued), and thought that the red tasted flat.

Leichhardt had learned from Helenus Scott that grapes at Glendon were all mixed together in winemaking, which the visitor thought most unlikely to produce good wine. He noted that the best course of action is to make wine from one variety; the grape that best suits the climate. He felt that grapes traditionally grown in Madeira, Tenerife or Spain would serve this purpose, but that perhaps French or German varieties would do equally well. 'Mr Scott told me that in France all the grapes are also mixed', mused Leichhardt, but 'I do not believe this'.[26] His observations emphasise the experimental state of selecting grapevine varieties to suit the Hunter, and the confusion over how to blend wines. In Bordeaux, wines were, and are still, made by blending portions of different grape varieties: Cabernet Sauvignon, Merlot and Petit Verdot or Cabernet Franc. However, this blending occurs after fermentation, when each wine's singular characteristics may be successfully combined with the others – not at the stage of fermentation.

After visiting a property near Singleton that did not have grapevines, but where the owner brewed beer, Leichhardt fancied that the brewing method – involving barrels, pipes and sealing of the fermentation container after the initial tumult of the yeast and sugar reaction – would be suitable for wine fermentation. Leichhardt seemed interested to try this closed fermentation at Glendon, where Helenus Scott allowed him to participate in the winemaking – fulfilling Leichhardt's curiosity about this mysterious process. The Glendon vintage began on 16 February, when the 'very ripe burgundy grapes' were picked (Leichhardt must actually mean the Red Hermitage or Shiraz, although Burgundy grapes are understood today to be Pinot Noir), and 'crushed with two rollers acting against each other'. The must from the crush was fermented in vats arranged in much the same way as those used by Parry and his vintage team at Tahlee. At Glendon, a 120 gallon cask with the base knocked out was set up in the vineyard as a vat, according to 'De Vonge's plan' (we cannot discern what is meant by this), and another vat was placed under the house. The must in the under-house vat began fermenting quickly, due to the warmer temperature.

On 18 February, the vat in the vineyard had not yet erupted into ferment. Leichhardt found the vintage enabled him to 'understand a number of things'. For example, he observed that rain appeared to slow the ferment in the open vat in the vineyard, which he attributed to rain containing nitric acid. He also witnessed 'drawing off' (the racking of the wine – or separation of fermented wine from the lees or after-products), 'purification' (fining for clarity) and 'sulphurisation' (the introduction of sulphur fumes to halt fermentation and prevent rapid oxidisation or spoilage). The size and quantity of the barrels at Glendon suggest but a modest quantity of grapes had been harvested.

On 19 February 1843, the Gouais, Sweetwater and Vantage (or Wantage, an English garden variety) grapes were picked at Glendon, 'cooled and brought to the casks' – and in a letter to a friend, Leichhardt described the processes of vintage at Glendon as poorly managed. 'Despite the expenditure they had made' – presumably on labour and equipment – 'nothing is in the correct place. We had endless trouble far into the night'. To avoid this, Leichhardt posited, 'a good arrangement, a bringing together of all casks in one place, would have made it child's play'. What is more, once the rain arrived on 18 February, Leichhardt predicted falls of at least a fortnight in duration as 'it seems to be that moderate widespread rain, which I observed on my arrival in Sydney'.[27]

Even so, Leichhardt was not standing apart as others made the wine – he participated actively, and with great delight. On 20 February, the naturalist noted in his diary that 'the blanket that I spread over the burgundy [grape] must tasted very sour while the underside was sweet'. As Leichhardt continued to record, 'Besides the carbonic acid, the warmth of the fermentation drives alcohol from the liquid as well, which is evaporated with the water vapour. As the alcohol contacts the air, it is changing into acetic acid combining with the carbonic acid of the air'. We should remind ourselves that for Leichhardt, and others engaged in the natural history studies that would later coalesce into specialised western sciences, winemaking processes were only just being revealed as chemical reactions between named acids and compounds.

There is one final feature of Leichhardt's experience of viticulture and winemaking that cannot escape notice. On 17 February – the day after the Scott family's vintage began – Leichhardt recorded that 'the Merton tribe, about 30 men strong, with old Gerry as king arrived in Glendon today'.[28] William Ogilvie produced wine at Merton for many years. Our research has yet to show that Aboriginal people harvested wine grapes at any Hunter property. Nevertheless, it is very likely that locals who remained in Country on settler properties in the region were involved in vintages at Merton, as well as at Dalwood, Camyr Allyn (Boydell's property in the Paterson Valley), and possibly even Dumaresq's St Helier's. In which case, Leichhardt's description of the arrival of this group of Aboriginal people is

Glendon in 1841. Album of cloud studies, mountain, bush and harbour scenes, c1841–50 drawn by Conrad Martens, f.70 Glendon. Unsigned. Dated 'May 5th, 41', State Library of NSW, Call no. DL PX 28.

important. The employment of seasonal labour for harvesting grapes and pruning vines is a window onto the layers of the community involved in – and attached to – Hunter wine: a community comprising not only vineyard and winery owners, but also the many hands required for the enterprise.

The Aboriginal visitors may have arrived from the Clarence River district in northern New South Wales, near where Ogilvie of Merton also held property, as they could not understand the language of Aboriginal people in the Hunter region. 'The young people were happy, playful, and many very well shaped', Leichhardt observed of the visitors recently arrived from Merton – and 'for the most part wrapped in possum cloaks', bearing long matching scars on their shoulders as marks of shared kinship.[29]

By 1843, patterns of cross-cultural contact in the Hunter region remained dynamic, with Aboriginal locals also reaching across boundaries of knowledge and practice, as did Leichhardt in observing and classifying, piecing together new patterns of understanding and being. If people like

Leichhardt were captivated by winemaking as science and ritual, for reasons apart from the production of alcohol, there is every reason to consider that Aboriginal people were also drawn to the engaging elements of the process.

When artist John Carmichael drew an Aboriginal family observing the pottery and vineyard at Irrawang, he captured the fact that Aboriginal locals, convicts and landowners intermingled in many places even after the increase in settler violence against Aboriginal people that is recorded as occurring from the late 1820s. Yet Aboriginal people were not on the edge of colonial estates, they were embedded in them, familiar with settlers and making their way as best they could through a rapidly changing physical and social landscape. By the early 1840s, fewer locals thrived in the region, according to accounts by Hunter clergymen who reported to the colonial government on the wellbeing of Aboriginal people.[30] This does not mean that they vanished, but Aboriginal people do largely disappear from our winegrowing story, because they moved away – or were forced away – from farms in the Hunter, or merged with the labouring community, or stopped identifying as Aboriginal.

A turning point in trade

IN 1833, BRITISH JOURNALIST CYRUS REDDING PUBLISHED ONE OF THE most influential books on wine to appear up until that time in the wider sphere of the British empire, *History of Modern Wines* (1833), including very brief mention of Australian wine. Redding wrote: 'it is true that which comes of science developes [sic] itself slowly, while the offspring of imagination careers in advance of time'.[31] The Hunter's first generation of winegrowers to 1846 were flamboyant personalities, fortune hunters of a sort; chaotic and obsessive imaginers who seemed almost to *will* Hunter wine into being by entrancing each other with their boundless 'careering' ahead. Between the 1820s and 1846 the Hunter became the agricultural heartland of New South Wales; most settlers who owned vineyards were intimates, or at least acquainted with each other, and they willingly co-operated as part of friendship groups and the wider sharing of interests. Family and social networks among winegrowers coalesced into a web that extended across the geography of the Hunter. Compared with the wider enterprise of settler colonialism and other types of commodity production, the strands of the winegrowing network were very fine, but strong, and individuals pushing forward with innovations benefited the winegrowing community.

As one committed to 'advancing' through trade, James King, in 1846 loaded five half-pipes (about a thousand gallons) of Irrawang wine at Morpeth wharf to be sent to Sydney. The *Shipping Gazette* reported with resounding applause that this wine was understood to be of superior quality and,

James King's wine label stamp, c1845, with a space to write the year the wine was made, under 'Vintage'. Newcastle Museum Collection 2014/55.

> we believe, the first parcel of that commodity that has yet arrived here regularly reported as merchandise at the Custom House. It is the commencement of a new and extensive trade with that important district, which will undoubtedly increase not only to partly supply Sydney and other southern parts of the colony with wine, but in time will form one of our staple exports – and one, too, which will not be confined to the English buyer alone, but will be a saleable commodity in every European colony or settlement to the eastward of the Cape … The Irrawang vineyard consists only of some six or seven acres of bearing vines, planted merely as an experiment; but the result of that experiment has been so highly favourable as to induce Mr. King very considerably to extend his plantation. His success and example too have stimulated others in the same quarter to plant the vine extensively, with the view to the profitable production of wine.[32]

Still, Hunter vinegrowers and winemakers were yet to conjure many drinkers of colonial wine from the dominant working class population of the Australian colonies, even in the metropolis of Sydney, where the middle class – the natural market for colonial wines – remained small and purchased imported wines of established taste and reputation. As Wyndham remarked to Leichhardt, most elite colonists preferred drinking tea to consuming colonial wine.

Chapter 4

Creating an industry through co-operation: the Hunter River Vineyard Association, 1847–76

DESPITE THE FIRST COLONIAL ECONOMIC BUST OF 1840 AND THE absence of an accessible market, ambitions for Hunter winegrowing did not falter. Supported by a circuitry of connections among the Hunter settler community, by the late 1840s the generation of imaginers had become a generation of Hunter winegrowing experimenters. The turning point came in 1847, when winegrowers from the wider agricultural community of the region broke away to form a more organised setting for co-operation, inaugurating the Hunter River Vineyard Association (HRVA). Here again timing proved to be an advantage for the Hunter wine community – the emergence of Hunter wine business occurring in parallel with developments in the wider worlds of colonial society, economy and empire. The HRVA, which pre-dated similar associations in the United States, became Australia's most enduring colonial organisation for winegrowing.[1]

The prevailing theme at the cusp of the generation of experimenters may be summed up by a remark in the Hunter Valley's first newspaper, the *Maitland Mercury* (itself a product of progressive settler colonialism in the region). Reporting on a Hunter River Agricultural Society meeting in 1843, a *Mercury* journalist mused that winegrowers were now 'more conversant with an art which has taken hundreds of years to bring to perfection in Europe' after the Roman invasion and the introduction of grapevines to Gaul.[2] Men of the British empire preferred early signs of progress, and some of the key figures in Hunter winegrowing were impatient for results – despite this realisation of the very slow pace of shaping expertise in vine cultivation and the even longer process of gaining a reputation for regional wines. As the rollcall of members of the Hunter winegrowing community grew, some such as James King and Henry Carmichael pushed against the pragmatics of the long game while others, among them John Wyndham and William Keene, persisted towards a further horizon.

Winegrowing progress in a bigger picture

THERE WERE SEVERAL PRECONDITIONS FOR THE FORMATION OF THE HRVA as New South Wales emerged from economic depression in 1843. Some of these factors were specific to the winegrowing community, some were specific to the Hunter Valley region, and others were of colony-wide import. In 1843, for example, licensing laws were extended from Maitland and Morpeth to Raymond Terrace in the Lower Hunter, to allow the sale of wine and beer by the gallon, which increased the viability of colonial wine production in the district where King, Carmichael and others were based.[3] Theoretically this provided a legitimate cash flow from wine sales. Yet most Hunter wine was still in all probability sitting unsold in cellars, or was disposed of in the common practice of part-payment of labourers' wages in farm produce. Nevertheless, now that quantities of experimental wine were increasing, winegrowers were focused on creating the conditions that would advance their ambitions to sell wine.

The opening in 1843 of the offices of the *Maitland Mercury* and *Hunter River General Advertiser* (1843–93), Australia's first newspaper outside of the Sydney region, can be understood as part of a wave of 19th century boosterism. That is, the escalation of public enthusiasm to increase economic progress through European-style farming and grazing in places perceived by settlers as wilderness or undeveloped wastelands. From the

1820s, such boosterism was typical in the English-speaking settler societies of Australia, the United States, New Zealand and South Africa in response to economic busts such as that of 1840. As historian James Belich asserts of this boosterist settlerism, uncultivated land seized from Indigenous populations provided economic opportunities for landless people from Britain and Europe. Many people of humble means were willing to leave Britain and Europe during the turmoil of industrial and political revolutions there in the 19th century, hoping for better lives through land ownership in new worlds.[4]

Books and newspapers were the principal instruments by which colonial improvement of uncultivated lands was spruiked and prescribed. While New South Wales newspapers were divided on many issues regarding civil rights and government policy, the press was united in a vision for economic progress, of which winegrowing in the Hunter was an ideologically significant, if materially modest, part. Moreover, newspapers were marketplace, meeting hall and (through re-publication of stories elsewhere) megaphone. With the inception of the *Maitland Mercury*, the story of Hunter winegrowing and the reputation of Hunter wine became amplified – colonially, intercolonially and internationally.

Also in 1843, among the 24 appointed members and 12 newly elected members of the New South Wales Legislative Council (not yet two houses of parliament), fully 11 men were either winegrowers or had close connections to winegrowing. The sole Hunter winegrower in this council was Richard Windeyer of Tomago. Laws enacted by the legislature in 1843 included the collection of statistics on vineyard acreages, total weight of grapes harvested, and the quantities of wine and brandy produced. This collection of data did not run smoothly at first, and the early figures are not a reliable measure of wine production for the Hunter.[5] The measure did however contribute to the long game of winegrowing by providing formal and quantifiable recognition of grape and wine production, and a basis for the later 'boosting' of grape growing for wine and table fruit to smallhold farmers throughout New South Wales.

Although structures were formalising to encourage winegrowing, not only did markets remain limited, but the lack of labour for established vineyards, and absence of interest among smallholders in small vineyards – and the lack of smallhold farmers *per se* – kept wine production at an experimental scale. To boost economic growth, the *Maitland Mercury*, for its part, supported crops of permanency in the Hunter, such as grapevines and olives, and backed a push for the new Legislative Council to foster an industrious class of tenants to rent small farms within large colonial estates, where such production might occur. Most large acreage Hunter properties purchased from 1821 had been virtually unworkable since the end of convict transportation in 1840 and the winding down of the assignment of those convicts yet to complete their sentences. Once convicts were no longer available as free labour, colonial productivity levels could not be raised without paying labourers wages, which eroded landowners' profits. Who would work large colonial estates? This led to campaigns, supported by colonists such as George Wyndham, to import poor indentured labourers from India and other British colonies. Were those humbler class colonists already living in New South Wales interested in renting land from James King, or from the Australian Agricultural Company, to earn a living as tenant farmers? Not, in 1843, to the extent that landowners or the *Maitland Mercury* envisaged. There were few attractions to responsibility of tenancy for the lower colonial ranks when labour was in short supply and high demand. The same large number of the colonial working classes who largely eschewed tenant farming also eschewed colonial wine, along with the still small colonial middle class.

A meeting of like minds

IN 1847, ENTRIES AT THE MAITLAND AGRICULTURAL SHOW INCLUDED a sample of red wine and a sample of vinegar from Andrew Lang at Dunmore; a red wine, a white wine and a vinegar from Montague Parnell at West Maitland; and a sample each of wines described as Burgundy, Sweetwater, brandy and vinegar (as well as table grapes) from William Charles Wentworth's Hunter vineyards. (Vinegar served many medicinal and cosmetic purposes in the 19th century, but that is another story.) The wine judges included Henry Carmichael.[6]

Hunter winegrowers belonged to a wider agricultural community, and at the Hunter River Agricultural Society dinner after the Maitland Show in 1847, discourse within this community progressed beyond incivility to shouting pitch, with some pastoralists and advocates of viticulture bitterly divided over how to fill in the land of the Hunter for economic gain. It became apparent that the great effervescent anticipation of settler fortunes to be made from land in Hunter's River had dissipated since the 1840 Depression. Yet – although native grasslands were manifestly not self-replenishing, and pastures in the Hunter now needed to be cultivated for grazing sheep and cattle – the ready export market for wool discouraged some landowners from divesting themselves of their sheep. Others – especially Carmichael and King – considered winegrowing to be the logical means to fill in the Hunter.

> On Wednesday, 18 May 1847, ten settlers met in the dining room of the Northumberland to inaugurate the Hunter River Vineyard Association.

Moreover, men without vineyards loathed winegrowers' capacity to legally (and no doubt appealingly) offer wine as part-payment to workers. There is no evidence that winegrowers were motivated to make wine only for the advantage of securing scarce labour, but wine did serve well as in-kind currency, and access to labour enabled continuing experiments with vineyards.

Following the heated post-show meeting in 1847, most Hunter winegrowers continued in their membership of the Hunter River Agricultural Society, and wine competition remained part of the society's show categories – but vineyard owners broke away to form the HRVA.[7] The investiture of this association signalled a turning towards more formalised sharing of knowledge and resources that had hitherto depended on family and social ties.

On Wednesday, 18 May 1847, ten settlers met by arrangement in the dining room of West Maitland's finest hotel, the Northumberland, to inaugurate the Hunter River Vineyard Association. A further two gentlemen who could not attend were also enrolled as members. King – one of those in attendance, and the principal agent of the push to formalise viticultural knowledge sharing – was elected inaugural association chair. WE Hawkins (about whom little is known) became honorary secretary, and standing committee members for the ensuing year were Carmichael, Andrew Lang and Edwin Hickey. Attendees at the lunch after the first meeting of the HRVA included John Morrison Saunders, manager of the West Maitland branch of the Bank of New South Wales. He proposed a toast to the vinegrowers of the district, to which Carmichael responded. Saunders' presence is evidence of the intimate and institutional relationship between the small professional class of the Hunter Valley at this time, and elite landholders, and as a winegrower he continued to attend the HRVA luncheon for the duration of the association and his career.

At the second meeting of the HRVA, in defiance of the wider regional mood against winegrowers part-paying labourers in produce, the matter arose of the profitability of winegrowing. The association meeting heard that investment returns from winegrowing were calculated as follows: from the *third year* vine-culture is capable of yielding a return adequate to cover the current expenditure involved in bringing its produce to market and the original investment will be repaid in the *seventh year* (emphasis in original).[8] The whip crack of italics in this *Maitland Mercury* report shows that passions were inflamed at the HRVA meeting against those settlers who accused winegrowers of *plotting* (our emphasis) to gain advantage in the labour market.

In fact, the calculation about return on investment for grapes and wine bore little resemblance to actual earnings among the experimenters. King admitted in correspondence to the Colonial Secretary in London in 1849 that after 16 years of vine cultivation at a cost of approximately £1000 to £1500 in total, plus labour, there remained no profit from the vineyard: 'my operations being on too limited a scale, and … necessarily mere matter of experiment, but experience has been gained'.[9]

From the third meeting of the HRVA in May 1848, the finer grains of the organisation began to emerge. Should members be obliged to bring their own wine to taste? Chair of the meeting, Archibald Windeyer of Kinross – a neighbour of King – held that all of the district's winegrowers should be entitled to attend, whether or not they had wine to present.

This sprang from the embryonic state of his winemaking, compared with his intentions. But where, asked Windeyer, could a settler learn how to foster his own investments in winegrowing, were it not through membership of the HRVA? A resolution was passed banning those unable to bring wines, and this seems to have become binding. Nonetheless, guests were invited to the wine tastings that occurred after the formal meeting ritual of the association, which exposed new producers and non-wine producers to Hunter wines. Some wines that had been tasted at the previous meeting were brought out from the Northumberland cellars, to see how they had changed in six months. Guests George Townshend and Alexander Park participated in the tasting, which included an imported Chambertin to compare with Hunter wines. We know Townshend did not make wine (and had been rendered landless in 1840). We also know Park's vineyard was extensive in 1853, but may have been in development in 1848. Lang brought wines made from Alexander Warren's vines at Brandon.[10]

From 1849, the HRVA lobbied to remove structural barriers to export from New South Wales to Britain. In the Legislative Council, William Charles Wentworth presented a petition from that year's HRVA president, Carmichael, protesting that British government customs duties treated Australian wine as foreign imports, while Cape wines received imperial preference. The problem, according to the petition, was that the home government, in Britain, was probably ignorant that wines could be readily and abundantly produced in New South Wales.[11] In reality, in the 1850s, there was little Hunter wine to send to Britain. Protests from the HRVA about British wine tariffs were largely symbolic; a way to begin to break into the tight web of mercantile politics of British wine trade.

Taste testing Hunter wines

ALTHOUGH THERE WERE FEW CUSTOMERS FOR HUNTER WINE IN 1847, HRVA members subjected their experimental wines to each other's scrutiny. And, through candid reports of their wine tastings in the *Maitland Mercury* – and re-publication of these reports across the colonies and abroad – the descriptions of the wines they presented to one another were broadcast widely, for the (nearly) 30 years that this long-lived organisation existed.

In 1850, the HRVA began blind tasting wines presented to the group – that is, giving attention to the qualities of the wine compared with previous vintages and contemporary competitors, without knowledge of the provenance of each wine. The origin of this practice is not stated in HRVA minutes, but it brought a more empirical approach, and reflected the method of judging at wine shows. HRVA members were also restricted to contributing two wines to each meeting. (Carmichael had brought seven to this meeting and decided to save the remainder for lunch.)

The blind tasting proceeded in the following manner. The association secretary numbered the wines, and removed them to another room. Once the tasting began, a hotel waiter served each wine by glass. Reds were tasted before whites, and the order of tasting within these categories was determined by the drawing of lots. When the *Maitland Mercury*'s Irish reporter John O'Kelly reported on the meeting, he ordered the wine descriptions by grower, grape varieties, style and remarks. At the 1850 meeting, King's Tinta, Black Pineau and Grey Pineau (Pinot Gris) blend received a high commendation from all present, and his 1846 Shepherd's Riesling (Semillon) was a credit to the maker.[12] Cory's 1850 Oporto (likely Tinta) seemed new although promising, but Cory's 1846 Madeira (Verdelho) was thin, tart and – according to Cory – ruined by unskilful management in casking. Lang's 1849 Red Hermitage (likely in fact a blend) was fruity and fine but left a strange aftertaste. Lang's 1849 white – made from Muscatel grapes – appeared strong, and indeed capital, and some at the meeting queried whether the wine was fortified with brandy, which Lang denied. Boydell's 1850 red – which he thought to have been made from a single varietal Shiraz – caused some excitement for tasting entirely of another grape variety, which led Boydell to conclude that his wines had been confused by his cellar hand. Boydell's Shepherd's Riesling earned a fair response, with Boydell excusing the wine as damaged by poor cooperage.

Carmichael presented an 1849 red blend of Pinot Noir wine and some other red wine made from a collection of grapes, being judged to be very strong and full flavoured, of Burgundy aroma. Carmichael's 1849

Shepherd's Riesling was considered one of the best wines of the meeting: luscious and most agreeable and, going around the table among the members: beautiful, delicious, sound, beautifully clear, and very good.[13]

Kelman's 1846 White Hermitage had improved even further on the excellence it displayed at the 1849 tasting. The bottle sampled had been sent to his brothers-in-law at Cassilis (presumably by dray) and then returned, to test how the wine withstood travel on the 200 mile round trip. A Kirkton Red Hermitage also met with great approval. Each of Kelman's wines was produced without the addition of brandy, and bottled without fining a year after manufacture.

The meeting tasted two further members' wines.

Windeyer from Kinross presented an 1849 red from Black Cluster grapes, pronounced as very good wine, very capital, and a magnificent wine. According to one taster, Windeyer's Shepherd's Riesling proved to be thinner than other Semillon samples present that year. Someone else decided the Windeyer white presented plenty of body, and a third held that this wine would improve in wood. Hickey's 1849 red wine, from Medoc or claret grape (Cabernet Sauvignon) was of pleasant flavour, but rather thin. Hickey blamed this lack of body on the cask being disturbed in his cellar. And Hickey's Hock – made from Madeira, but which he was inclined to think was Pineau – was poor quality from picking the grapes too early in heavy rain, leaving the wine unripe, and thin, and not matured. Some brandies were also tasted over lunch.

Three other matters arose at the 1850 meeting. First, the arrival of German emigrants recruited through the colonial bounty scheme managed by Wilhelm Kirchner (discussed further in the next chapter). Second, Carmichael's intention to lead the Hunter's submissions to the 1851 London Exhibition at the Crystal Palace and, third, James King's presentation of correspondence on wine science from German scientist Justus von Liebig. Interest in the London Exhibition and King's correspondence with Liebig (comparing Hunter wines with German products) were significant initiatives by these core HRVA members.

During the economic recovery beyond the Australian colonies, a melding of the forces of industrial modernity in manufacturing, transport and communications was signified most powerfully with the staging of the Great Exhibition of London, held in the purpose-built Crystal Palace in 1851. This was the first of many international exhibitions of the modern age, and incorporated a wine show.

Carmichael did not become the New South Wales agent for the London Exhibition – that role went to William Macarthur. What is more, Hunter wines were *not* sent to London in 1851. This infuriated Carmichael, who had especially prepared a selection of wines made by King from grapes grown at Porphyry Point. Carmichael laid the blame for this blunder in several directions within the colonial administration, leading to a curious series of events.[14] Carmichael sold his wines that had been intended for London to a colonial buyer, and then bought the bottles back from the buyer, in order to ship them to Buckingham Palace, to gain royal attention for his wines. This bold manoeuvre resulted in an encouraging missive from the Palace about the quality of the wines for which King, as the winemaker, gained quite some colonial glory.

> For Hunter winegrowers, the London Exhibition precipitated a rapid scaling-up in consciousness from local to global opportunities for showcasing their wines.

For Hunter winegrowers, the London Exhibition – and the subsequent age of international exhibitions – precipitated a rapid scaling-up in consciousness from local to global opportunities for showcasing their wines. Hunter wines were also entered in the colonial competition of the Agricultural Society of New South Wales, in Sydney, and in intercolonial events.

King's correspondence with Liebig staked some territory for the reputation of the Hunter as being at the forefront of the extension of winelands to the antipodes alongside the efforts of William Macarthur, which benefited from Macarthur's extensive family, social and professional network in Britain and Europe.

After the second HRVA meeting of 1850, in November, the *Maitland Mercury* published a report from Henry Lindeman on his wines from Cawarra. Lindeman presented an 1849 red Cyras fermented in open vats on grape skins before being casked, and then racked four times after fermentation to remove the lees. Lindeman's white wine from Roussette grapes (possibly grown from Kelman's vine stocks, but called Roussanne) was made in the same manner, except that, Lindeman noted, grape skins were excluded from the white.[15] At this stage Hunter winegrowers were not all aware that white wine was made without grape skins.

Brethren of the pruning knife, godfathers of the vine

MANY WINEGROWERS EXCHANGED VINES RATHER THAN PURCHASING identified stocks. For this reason, confusion had arisen in the Hunter region over the names of vine cultivars, and therefore the wines that could be expected to be made from them. Between the two 1850 meetings of the HRVA, an argument erupted over this matter in the pages of the *Maitland Mercury*. The debate opened with an unnamed gentleman appealing to his 'brethren of the pruning knife, the godfathers to their own vines'. He called for greater precision in identifying varieties presently planted in vineyards, as well as during distribution of vine stocks in the future, as many old hands were extending their vineyards, and some new investors joining the brethren.[16] This gentleman thought the Black Spanish or Lambrusquat in his vineyard might be from a Paterson vineyard created from vines obtained from Busby's vine collection. Whether he means the 1830 Busby distribution of vines at the Botanic Gardens, or the later collection from Europe is not clear – and the conflation of these two collections has led to confusion. The gentleman stated, further, that Kelman's White Hermitage was in fact Marsanne not Roussette, as White Hermitage was made from two grapes, Marsanne and Roussette or Roussanne, and Kelman made single varietal wines. (Intriguingly, a white grape variety was known to be planted at Maria Windeyer's Tomago, Archibald Windeyer's Kinross and also at Windermere that could have been either Marsanne or Roussanne. Tomago Estate also grew Black Cluster grapes, as Windeyer's son William asked after their ripeness when he wrote home from Sydney about the vintage of 1851.)[17]

While the original correspondent in the 1850 debate believed there to be no white Scyras (Roussanne) in the region, another correspondent wrote that this grape had been grown in the Hunter for at least 20 years – and well before Busby's European collection of vines was developed at Kirkton.[18]

This matter remained unresolved.

Maria Windeyer, Tomago (Tomago)

NOT ALL WINEGROWERS IN THE REGION AT THIS TIME WERE MEMBERS OF the HRVA, nor were they all brethren. At Tomago, Maria Windeyer depended on income from wine, but was probably excluded from the HRVA simply because of her sex. Windeyer counted upon her wine to support not only herself but also one of her sisters, whose husband was permanently hospitalised, as well as her sister's children.

The acute nature of Windeyer's financial circumstances is revealed in letters she received from her son William during his schooling in Sydney, when he handled the freight of samples of Tomago wine to Sydney, Adelaide and London.[19] Unfortunately, in 1849, former French *negociant* (wine merchant) Didier Joubert in Sydney regretted that he could not sell Tomago wine, although it seemed sound (but could be more carefully fined). Joubert had quite enough colonial wine to hand (presumably from William Macarthur and *speculatively* from James King), and there remained a strong prejudice against colonial wines.[20] Another merchant found Tomago whites a little too sharp for sale, but thought the reds to be saleable.[21] Nonetheless, in 1850, the Australasian Botanical and Horticultural Society awarded Maria Windeyer a gold medal for her wine, at which event William Burnett (who advised the *Maitland Mercury*) also received a gold medal for his champagne.[22] This may have propelled Tomago wines into greater favour among merchants, but this is not clear from the further letters between Maria and William.

William Windeyer welcomed the discovery of gold at Bathurst in New South Wales in 1851 – the ignition of the Australian gold rushes – as a 'good thing for our servants and a good thing for our wine'. William remarked too that many of his school fellows were being called home to their family's pastoral properties to work, 'because all of the shepherds are running away' to the goldfields.[23] Colonial wine sales in New South Wales were indeed boosted by sales at stores established to provide for the diggers for the following decade, but more significant income from wine and grapes would be made on the Victorian goldfields, which benefited Victoria wine producers, and those in the Riverina at the border of Victoria and New South Wales.

Maria Windeyer's wine received a certificate of commendation at the 1855 Paris Exhibition, an achievement that is all the more remarkable for the challenges she faced as a woman selling wine into the male-dominated world of the Sydney wine market.

And there are further members of the Hunter wine community to introduce for this era.

John Glennie and Thomas Patch, Orindinna (Gresford)

JOHN GLENNIE WAS BORN IN ENGLAND AND EDUCATED IN SCOTLAND, receiving his Master of Arts from the University of Aberdeen in 1835, before immigrating to New South Wales in 1836. He followed other family members to the colony, and after working in the central west had gained a position as manager at Glendon by 1843. Later in the 1840s, Glennie invested in property at Wee Waa and in the Paterson Valley estate Orindinna, originally occupied by his business partner Thomas Patch and then leased by John Lankester.

Patch, an English-born medical doctor, was employed by the British East India Company, in India, before immigrating to New South Wales. In 1841, Patch purchased Orindinna from George Townshend's insolvent estate. The Lindeman family resided with Patch until they relocated to Cawarra, and Patch was friends with the Scott brothers, Alexander Park and Charles Boydell. Patch appears to have planted grapevines in 1844. He recorded some success making wines from Verdelho, Shepherd's Riesling and Shiraz grapes, and participated in the early years of the HRVA. Patch and his wife returned to England in the mid-1850s, and Lankester leased Orindinna before going on to become a key figure in the Riverina.

In the late 1850s, Glennie sailed to England, married, and returned to live at Orindinna, focusing on winegrowing – to international acclaim.

Glennie won a gold medal at the Paris Exhibition of 1878, but for general sale of his wine Glennie relied on John Wyndham's channels of wine distribution. Wyndham – son of George Wyndham – was married to a daughter of one of Glennie's cousins. Glennie took a central role in the HRVA from 1861, and it is noteworthy that the record of his tenure at Orindinna is replete with experiences of floods and hail, although he continued his successes through to his death in the 1870s.[24] Glennie also became a principal wine judge for the Maitland Show.

Becoming an industry

BETWEEN 1847 AND 1876, HUNTER WINEGROWING BECAME AN INDUSTRY; a group of mainly family firms in competition for a market that was vertically integrated. In other words, these family owned and operated businesses grew grapes (and often bought in other grapes), made wine *and* sold the wine through their own retail outlets as well as wholesale to other sellers. As explored by historian James Simpson, the family wine business model of vertical integration in the Australian colonies was distinct from traditions of horizontal integration in the British wine trade, whereby British merchants were installed in the port cities of Bordeaux, Madeira, Oporto and Jerez, providing the link between grape growers in wine countries and wine buyers in Britain (and internationally).[25]

In the Hunter, the vertical integration of family wine businesses (along what is known as the wine chain from producer to consumer) bypassed the tradition of British wine merchants, who linked drinkers with European styles through stories of wine quality that influenced consumer preference. Hunter winegrowers sold wines from their wineries in small quantities (not more than 2 gallons at a time) and also through direct postal orders from drinkers, sales to general stores, sales to publicans and wine shop owners, and – in the case of the Lindeman and Wyndham families – through wine cellars that they owned in Sydney's Pitt Street from the 1860s. The Pitt Street wine cellars aimed at a market of well-to-do urbanites who drank at home in their daily routines, and frequently entertained guests.

Wine shops were a new way of selling colonial wine created by the New South Wales *Sale of Colonial Wine (Cider and Perry) Act* of 1862.[26] These were places for the working classes to drink wine on the premises. The shops were located in the front rooms of humble houses on main roads through the bush, in Sydney, Newcastle and smaller towns. As it happens, John Wyndham frequently received requests from people of the lower colonial ranks asking him to pay the modest wine shop licence fee, in exchange for selling his wines exclusively. At first this licence cost only £1, compared with £30 for a publican's licence, to encourage more wine shops (in the 1880s it rose to £10, but even into the 20th century the fee remained much lower than a publican's licence). Wine shops carried low overheads, and served a humble custom. It is evident from Wyndham's correspondence that the operators of these establishments were often of limited literacy. The opportunity to open a wine shop presented them with a chance to improve their circumstances by obtaining a small land selection (in rural areas) and operating a business. It is difficult to determine how widespread this practice of winegrower-paid licences was. Yet Wyndham received letters from further afield than the Hunter to sell his wine in this way, revealing a little-known culture linking winegrowers with humble colonial business people and drinkers; albeit negligible compared with the dominance of pubs selling beer to working people.[27]

> Wine shops were a new way of selling colonial wine created by the New South Wales *Sale of Colonial Wine (Cider and Perry) Act*.

To reach the market of people for whom drinking wine existed as part of their wider expression of good taste, the Hunter had to develop a reputation for distinctive wines. In the long game of wine business as developed by the British wine trade, such reputations required attention to winemaking practices and also to the linking of wine styles with localities (but not single vineyards) and leading brands, through storytelling. In Europe, ordinary wines always existed alongside those wines elevated to special status – otherwise the common people would have no affordable wines – but to understand this required experience of travelling in European wine countries.

William Keene, modernity and locality

PRIOR TO IMMIGRATING TO NEW SOUTH WALES, ENGLISH GEOLOGIST William Keene worked for many years in France, where his extensive travels provided him with a comprehensive knowledge of French grape and wine cultivation, gained through observation of the differences in regional wines and vineyard landscapes. Keene worked near Bordeaux and learned to dress vines at his vineyard in the Pyrenees, applying the same scientific curiosity to this process as he did to his profession as a mineralogist. Keene arrived in Sydney in 1852, and by 1854 was appointed the New South Wales examiner of coalfields, located in the Hunter from 1863.[28]

Keene is remembered for encouraging the use of saccharometers in Australian wine manufacture. Although John Wyndham already used such a device at Dalwood, Keene refined and broadened this application. A saccharometer is a type of hydrometer that measures the specific gravity or mass of sugars in grape must, to calculate Brix (see Introduction). Winemakers in this era also estimated sugar–acid balance in ripe grapes before harvest, as the quantity of grape sugars to be converted into alcohol must be balanced with the grape acid to result in a sound wine. Too much *acid* will result in insufficient fermentation to convert all grape sugars into alcohol. Too much *sugar* may leave grape sugars unfermented, and create a sweeter (or flabbier) wine than desired. In Keene's discussion of saccharometers with the HRVA in 1865, it emerged that Wyndham had bought a device from a London company in 1857. No such equipment could be procured in Sydney, although devices for brewers were readily available. Keene confirmed Wyndham's prudence in not simply adapting brewing equipment, by demonstrating that grape must contains a higher sugar content than brewed beverages, requiring a different type of instrument than one used in the making of beer or ale.[29]

Keene also presented Hunter winegrowers with a view of wine culture as a relationship between the natural conditions of a place and the cultivation of grapevines, influenced by his experience in France. In his 1865 presidential address to the HRVA, Keene compared the subtleties of French winegrowing practices from the north of the country to the south.

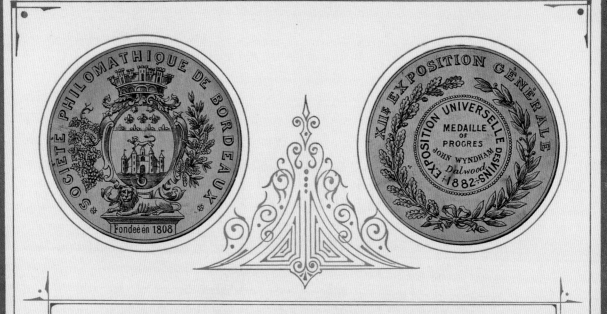

BORDEAUX INTERNATIONAL EXHIBITION.

AWARDS. "By direction of the wine Jury, the name of **JOHN WYNDHAM** was ordered to be placed first (1st) among all wine exhibitors from the Australian Colonies."

AMSTERDAM INTERNATIONAL EXHIBITION.

AWARDS. Diploma of honor to John Wyndham, Esq. of Dalwood, for collection of red wines, Hermitage & Verdot, which are the most neuter & the finest of both Colonies (N.S. Wales and Victoria)."

Some of John Wyndham's many wine awards. H.G. Ballard, Photographs of the Dalwood Vineyards near Branxton, New South Wales, Australia, 1886, State Library of NSW, Call no. PXD 740.

He explained the variance in trellising of vines (height of vine trunks and canopy management) to take best advantage of light and warmth. Keene explained that across France's considerable expanse of territory, different grape varieties were cultivated for wine, vinegar and eating fresh. He painted a picture of wine as a nationally coherent form of cultivation – something that had been lacking in the conversations within the Hunter winegrowing community – paying eloquent attention to the identities of wines in various locations and the modest drinking habits of the French arising from their lives entwined with grape and wine production.[30]

Keene recommended that his peers shape Hunter vine cultivation to suit Hunter conditions. He urged them to put aside unhelpful advice (remarking that most of the advice they had received to date could be categorised as unhelpful), and to direct their efforts to responding to their local environmental conditions and to specific industry needs to strengthen the region's winegrowing culture.[31] France's traditions of grape growing and processing were a response to conditions there – Keene explained – an exemplar, certainly, but not automatically a blueprint for other wine countries such as the Australian colonies.

We know little of Keene's wines, but his influence on Hunter winegrowing practices during his membership of the HRVA, and as president from 1865–67, marked a significant turning point towards greater insight into the most suitable sites in the region to plant *V. vinifera*.

Keene is buried at the Anglican cemetery in Raymond Terrace and memorialised in the town's Anglican church.

Philobert Terrier, Kaludah (Lochinvar)

KALUDAH VINEYARD AND WINERY AT LOCHINVAR ORIGINATED AS A component of a broader portfolio of North British Company (NBC) investment in land development in the Hunter Valley from 1840. The NBC was a joint stock company, known as the Scotch company, and in competition with the Scottish Australian Investment Company (SAIC) – although officers of the SAIC appear to have encouraged the NBC to purchase land in the Hunter Valley. In the 1830s, Scottish settler capitalists associated with the SAIC had sponsored many hundreds of Scottish labourers to the colony of New South Wales. Lang of Dunmore alone sponsored 253 immigrants, by no means the largest figure, but still a substantial quantity of manpower to support settlers in both their need for labour and their hopes for a tenant base to bring about agricultural development. In 1845, the manager at Kaludah, James Blake, replanted 33 acres of vineyard above the high-water mark from flooding – although other reports dispute the extent of this vineyard.[32] NBC lots were sold to John Frederick Doyle in 1859.[33]

Despite the tremendous influence of the idea of French winegrowing culture in colonial New South Wales, few French people actually immigrated to the colony, and therefore direct French influence in the Hunter is rare. In 1855, however, Frenchman Philobert Terrier was employed (presumably by the NBC) to manage Kaludah vineyards and winery. In 1867, Terrier expressed satisfaction in the quality of wine produced at Kaludah after favourable vintage conditions. He wondered however about limitations on selling wine in bulk from wineries, as occurred in France, which he considered a great impediment to the ready circulation of products in the colony of New South Wales. Terrier also raised the issue of climate parity with Madeira, suggesting that since the vineyards there had been all but destroyed by powdery mildew and the depletion of soil fertility, the Hunter seemed well positioned to replace wines known as Malvasia, Tinta and Dry Madeira in the global marketplace.[34]

Powdery mildew

AS POWDERY MILDEW DEVASTATED WINELANDS SUCH AS MADEIRA, THE outbreak of the disease in Australia threatened to nip Hunter winegrowing in the bud. Powdery mildew (*Oidium tuckerii*) first came to colonial Australian attention among grapes in Queensland in the 1860s. The destructive capacity of this white mildew – an invasive disease that arrived from North America's native grape environment, via Europe's *V. vinifera* industry – incited a frenzy of concern, particularly within the HRVA.[35] Winegrowers called for severe penalties for vine stock transfers from infected areas. Fortunately, a solution was quickly found in the application of sulphur spray (applied with bellows in the colonial era). The arrival of powdery mildew counts among the tremendous shocks to wine industry development in Australia that have faded from national memory. (Hunter grapevines are not infected with phylloxera, but have suffered from downy mildew from the 1920s, as discussed in chapter 7.)

Joseph Broadbent Holmes, The Wilderness and Caerphilly, Allandale (Rothbury)

ENGLISHMAN JOSEPH BROADBENT HOLMES AND HIS WIFE IMMIGRATED TO New South Wales in 1843, and purchased a portion of a land grant at Allandale, between Maitland and the contemporary wine district of Pokolbin. According to descendant Jim Fitz-Gerald, Holmes acquired vine stock from Dalwood, and by 1844 had planted out a vineyard at The

Joseph Broadbent Holmes, The Wilderness. Sketch by Marianne Glennie, c1867, copy held at The Edgeworth David Museum, Kurri Kurri.

Wilderness. The family also purchased a property called Caerphilly at Rothbury, and Holmes's sons became second settler generation winegrowers.[36] Born in 1850, at Allandale, Charles Phillip Holmes succeeded his father at the head of JB Holmes & Company, serving as the chairman of the board of the family company and the Hunter Valley Distillery. His obituary suggests he maintained these roles until his death, aged 78 years.[37] The Holmes family contributed land and funds to build St Paul's Church of England, Rothbury, on the road between Branxton and Pokolbin. Before the construction of this church, led by the Reverend Alfred Glennie (who is buried in the churchyard), Glennie and Reverend Lovick Tyrrell (brother of winegrower Edward Tyrrell) preached in Joseph Holmes's house – evidence of the natural entanglement of winegrowing and worship in the Hunter wine community.

Fitz-Gerald's father's memories of The Wilderness included meeting an elderly labourer who had served as a drummer boy at the battle of Waterloo, who may perhaps have been known to Henry Dumaresq from the generation of wine imaginers. Another of Fitz-Gerald's relatives related the story of visiting Holmes's properties after the vintage, and watching horses and cattle fed the wine skins racked off after the grape ferment: the sight of the animals lying on their backs waving their legs in the air blind drunk was hilarious.[38]

A dedication ceremony at St Paul's church in the late 1930s reunited members of the key winegrowing families of the region, and indicated how marriage over subsequent generations had entwined earlier settler generations of Holmes and Glennies, and different branches of the Tyrrell family, including those at Ashmans in the heart of Pokolbin.

Edward Tyrrell, Ashmans (Pokolbin)

ENGLISH BROTHERS LOVICK AND EDWARD TYRRELL EMIGRATED FROM KENT to New South Wales in 1854 to join their uncle William Tyrrell, who served as bishop of Newcastle diocese, based at Morpeth (when this township was still the centre of Hunter society and economy), from 1847 to 1879. Lovick was vicar of St Peter's, East Maitland, for 40 years, and preached in the surrounding district. In 1858, Edward Tyrrell purchased 320 acres of land at Pokolbin, and named the property Ashmans, for his grandmother's ancestral home in Suffolk, England. A report on proposed roads in Pokolbin in 1875 refers to EJ Tyrrell's 320 acres as a landmark.[39] Edward Tyrrell was a Justice of the Peace and, according to Tyrrell family history, built his winery in 1863 and made his first vintage in 1864. By 1870, the 30 acre vineyard contained Shiraz, Semillon and Aucerot vines. In 1869, Edward married Susan Hungerford. Among their ten children, Edward George Young Tyrrell (known as Dan), born in 1871, began making wine in 1889.

Joseph Drayton, Bellevue (Pokolbin)

ENGLISH-BORN JOSEPH DRAYTON, HIS WIFE ANNA AND TWO SONS SAILED from Britain to New South Wales in 1852, when Joseph was aged 27 years. Tragically, Anna died during the voyage, along with a newborn daughter and one of the Drayton sons. Joseph and his surviving son Frederick arrived at Lochinvar in 1853. After some years, Joseph acquired 80 acres of land at Pokolbin and constructed Bellevue homestead. (Frederick later became a blacksmith.) Joseph and his second wife Mary Ann Chick had eight children; their son William joined his parents as a Pokolbin landholder, planting vines on a 40 acre selection of land. William and his wife Susan (nee Lambkin) divided their estate between their ten children, resulting in the dispersal of Drayton property ownership across the blocks named Bellevue, Happy Valley and Crow's Nest.

Alexander Munro, Bebeah (Singleton)

IN INVERNESS, SCOTLAND, LATE ONE NIGHT IN MAY 1829, 14-YEAR-OLD Alexander Munro and two other boys robbed a grocery story of money and goods. The next morning, as Alexander slept, his mother Isobel (a widowed domestic worker) found coins from the robbery and reported the theft to the police. One of the other two boys confessed to the crime and earned a lighter punishment, but Munro and Simon McKenzie were arrested, tried and sentenced to transportation to New South Wales for seven years. The pair sailed to Sydney separately in 1830, arriving in 1831. According to family historian Jillian Oppenheimer, Munro was fortunate to be assigned immediately as a labourer to John Browne, an emancipist settler formerly of the Hawkesbury, who had been drawn to the better farmland of Patrick's Plains in the 1820s.[40] The Hunter's community of former Hawkesbury settlers was permitted convict labourers, although it was allowed fewer assignees than the settler capitalists who dominated the earlier period of Hunter winegrowing history. Browne's land allotment adjoined that of Benjamin Singleton (for whom Singleton is named).

Munro worked for Browne until he completed his sentence in 1836, after which Browne assisted him in purchasing a team of bullocks and running a carrying business between Singleton and the Morpeth wharf. With capital accumulated from five years in this business, Munro worked in various jobs as a butcher, shearer, storekeeper and (reputedly quite a fine) baker. In 1838, he married Sophia Lovell, a convict assigned to Francis Little at Invermien. Between 1837 and 1842, Munro bought land and businesses from Singleton, who was declared insolvent during the 1840 Depression – making Munro another beneficiary of the reordering of colonial land ownership and society that admitted Lindeman and some others to strategically located properties. Alexander and Sophia invested in hotels and vast inland pastoral runs (benefiting from long-term leases following legislation in 1847). Munro became the first mayor of Singleton, and was a leader of the Masonic community and of district agricultural societies. Perhaps because he and Sophia were childless, they became remarkably generous benefactors of Munro's relatives in Scotland, and encouraged the migration of kinsmen and women from Scotland to New South Wales.

In 1860, Munro established Bebeah vineyards and later purchased other vine properties. He situated Ardersier House (named for his place of birth) in view of Bebeah vineyard, and outfitted the winery with a state-of-the-art winemaking plant, costing £7000. Kinsman William McKenzie

> Alexander Munro and Simon McKenzie were sentenced to transportation for seven years. The pair sailed to Sydney separately in 1830, arriving in 1831.

managed Munro's wine enterprise.⁴¹ In the previous century, wealthy merchants in Scotland had ornamented their homes with vines to signify civilised wealth, and filled their cellars with great European wines. Unlike them, expatriate Scotsman Alexander Munro completed his days gazing upon his own vines and (presumably) drinking his own wine.

An industry for empire

AS MUNRO ENTERED A RETIREMENT OF ARCADIAN EASE, JOHN WYNDHAM bore the heavy burden of managing the Wyndham empire of vines at Dalwood, and also at his brother's property Fern Hill, and at Bukkulla at Inverell – among other productions. John's brother Hugh Wyndham, who was based in the New England region, wrote to his sibling in 1869 to report on the wool clip at Bukkulla and other inland properties. Hugh expressed concern that John worked very long hours: 'I was very sorry to hear you have been unwell, but surely you should rather lay it to the 14 hours a day office work than to the great heat – [the] latter would not hurt you but 14 or 15 hours a day in an office !!! who could stand that?'.⁴²

Also in 1869, John Glennie urged Wyndham into a more focused competitiveness against Albury wine baron James Fallon. Glennie felt certain that Fallon's enterprise could not be as impressive as Dalwood, and Glennie intended to demonstrate this to a member of the New South Wales judiciary who would soon be visiting. As Glennie continued in his letter, Fallon may have *splendid* 1600 gallon casks, but 'he is beaten in this when [the visiting magistrate] sees your 3500 ones!'. Glennie hoped for intercolonial free trade in wine, so that 'we can fight Fallon and all the rest of them'.⁴³ Hunter rivalry with Albury united some Hunter winegrowers against a common opponent.

At least as early as 1870, Wyndham of Dalwood wines from Bukkulla were sold in London, by the merchant David Cohen & Co.⁴⁴ Wyndham dealt with the Maitland branch of the English company, a connection that surely fostered this sale of Bukkulla wines at the centre of the web of imperial trade. The quality of Bukkulla wines in 1870 may be credited to Frederick Wilkinson, who managed this vineyard and winery for

Wyndham in the late 1860s, as his own new vineyard at Pokolbin reached bearing. Among the Wyndham records, Wilkinson's knowledge of and skill in vinegrowing and winemaking fairly crackles from the pages of his correspondence with John Wyndham. Meanwhile, Wyndham's agent in Newcastle found that in this emerging port town – teeming with seamen and the working classes of surrounding coal mining towns – Wyndham's Dalwood wines were in large part preferred to Wyndham's Bukkulla.[45]

Rivers of strange wine

IMPERIAL TRADE PREFERENCES WERE CHANGING. BRITAIN RECEIVED RAW wool for processing from Australia, and so no longer needed wool from Portugal – a trade dependency that had prompted the wool-and-wine trade terms of British preference for Portuguese wines since the early 18th century. In the 1860s, British authorities began to favour lighter-bodied French wine imports, through tariff preferences (which would have greatly cheered philosophers Adam Smith and David Hume of a century earlier). Fashions for alcohol in the western world tend to ebb and flow over long periods of time between phases of high alcohol consumption leading to social problems, followed by discouragement of inebriety and the encouragement instead of greater civility in drinking. (Also, by the 1880s, advances in medical science led to a redefinition of drunkenness as an addiction, or a disease of the body, rather than as a moral flaw to be criminalised – although drunkenness was not yet formally decriminalised in Britain or the Australian colonies.)[46]

In this new era of empire health and wine politics – in both trade and taste – the combined scientific and moral authority of medical doctors drew them to the fore as wine commentators. Henry Lindeman's prominent role in New South Wales campaigns of the 1860s to encourage the consumption of inexpensive wine as opposed to other forms of alcohol was influenced by British medico Robert Druitt.

In what were perhaps the earliest British publications linking wine and health, Druitt's newspaper columns about the best of 'cheap wines' were a response to 'the rivers of strange wines' that were 'coming in from all parts

John Wyndham. Photographer William Bradley, State Library of NSW, Call no. P1/2007.

of the world', because 'both the medical profession and the public wanted to know what they were good for'.[47] Druitt's convictions were as clear as the wines that he pronounced to be most healthful. 'Civilised man must drink, will drink, and ought to drink; but it should be wine', he wrote. According to Druitt, the moral properties of wine compared with other forms of alcohol lay in wine's qualities as a nutrient. He also claimed that as long as wines were from famous places, even the most common (cheap) of them contained mysterious flavour variations that would also mentally transport and beneficially transform the drinker. 'Alcohol is a drug', wrote Druitt, 'that could be purchased as cheaply as 2 shillings a gallon'. Although wine also contained alcohol, he continued, *what the physician or connoisseur valued in wine* was that 'One mouthful of Madeira or Burgundy will produce an instant exhilaration, because of the wine flavours, which no quantity of sheer alcohol and water could produce'.[48]

Among the strange rivers of wine Druitt found flowing into Britain was the Bukkulla red made by Fred Wilkinson, available at David Cohen & Co, which the good doctor described as a stout, admirable wine with peculiarly shaped crystals.[49] Druitt reported too that a recently visiting Frenchman had declared the white wines of Bukkulla to have exquisite finish.[50] Perhaps Henry Lindeman, who read Druitt, mentioned to Fred Wilkinson at an HRVA meeting that Bukkulla wine had received Druitt's praise? Lindeman, Wilkinson (and his brother John Wilkinson), were all members of the HRVA in the 1860s – an association by this stage comprising a new generation of characters different from those who began the group two decades earlier.

The Wilkinson family

IN 1852, ALFRED WILKINSON AND HIS FAMILY ARRIVED IN VICTORIA to rejoin the military regiment in which he had served in India for the British East India Company. His posting in Victoria until 1859 places Wilkinson in the midst of the southern colony's gold rushes. The record becomes less clear for several years until, from 1866, Alfred's son Charles Wilkinson is recorded as the landholder of Oakdale at Pokolbin, a 98 acre portion of land that the family later increased in increments. In Willkinson family lore, Alfred's sons Fred and John (and perhaps William) underwent training in Europe to prepare them for careers in vinegrowing and winemaking: Fred with the influential German chemist Justus von Liebig (who may have been in England at this time), and John with Monsieur Lanseur of Fleury, about which no more is known. Fred Wilkinson planted Oakdale's vineyard and was employed from 1866 to 1870 as vineyard manager and winemaker at the Wyndham family's Bukkulla Estate at Inverell.[51]

In 1875, Fred Wilkinson married Florence Mary Hindmarsh Stephen, granddaughter of a governor of South Australia, George Milner Stephen, by all accounts a brilliant man whose ambition seems to have extended his pursuits into colourful territory, including that of faith healer.[52] Florence and brother Lionel's family connections go some way to explaining the

Wilkinson family at Maluna, c1910s. Courtesy of Don Seton Wilkinson and family.

Maluna, c1910s. Courtesy of Don Seton Wilkinson and family.

great care taken by Fred and Florence's son Audrey Wilkinson in the manner of his recordkeeping for Oakdale and other Wilkinson properties, from the beginning of his extant seven-decade-long diary series, and his famous gentlemanly manners.[53]

John Wilkinson married Jane Hungerford (a relative of Susan Tyrrell, mother of Dan Tyrrell), one of many Hunter Valley descendants of Emmanuel and Catherine Hungerford, settler capitalist emigrants from Ireland, who arrived in New South Wales in 1828 and bought property at Wallis Plains. William Wilkinson managed Kelman's Kirkton vineyard and winemaking at a time when these wines were receiving considerable acclaim – with wine exhibited and sold under the name James Kelman, the second generation settler landholder. William Wilkinson's employment at Kirkton goes some way to explaining how vine cuttings from the Kelman vineyard may have been significant in Pokobin, where the Wilkinsons' properties were clustered. This cannot however be considered cut and dried without some recourse to community memories about vine transfers, as vine cuttings were equally likely to have been obtained from Dalwood as a result of Fred Wilkinson's connections with John Wyndham, or indeed, from elsewhere. The Wilkinson properties were Oakdale, Côte d'Or, Maluna and Coolalta. The history of this family's employment formed an important bridge between the Wyndham and Kelman enterprises and the evolution of embedded expertise at Pokolbin during the subsequent generation of the Hunter winegrowing community.

The HRVA, a purpose served

NEW ECONOMIC PROSPERITY AND SOCIAL OPTIMISM IN BRITAIN FROM THE 1820s saw the emergence of the first modern British wine writers, Cyrus Redding and, before him, Alexander Henderson. A Scottish medical doctor, Henderson's *The History of Ancient and Modern Wines* appeared in England in 1824 and was available in New South Wales in 1828.

Henderson described the pleasure to be gained from knowledge of the provenance and 'heroes' of wine in classical times, and the present. He wrote that:

> The invention of wine, like the origin of many other important arts, is enveloped in the obscurity of the earliest ages: but, in the history of ancient nations, it has generally been ascribed to those heroes who contributed most to civilise their respective countries, and to whom divine honours were often rendered, in return for the benefits which they had conferred upon mankind.[54]

As Henderson's words were being read by wine aficionados in British society, James Busby's first book, *A Treatise* (1825) appeared in Sydney with the purpose of seeding New South Wales winegrowing (not realising this was already underway), which did earn him the very reputation of heroism to which Henderson alluded. As we argued in the Introduction, Busby's story overshadows other characters of longer and more sustained contribution. One such character was *Maitland Mercury* reporter John O'Kelly, who attended HRVA meetings for 20 years, and whose reports provide us with the backbone of the unfolding story of the Hunter winegrowing community.

The death in 1867 of John O'Kelly must therefore have been felt as keenly by the HRVA as it was by the wider Maitland community, who mourned him greatly. Indeed, the final demise of the association occurred in 1876, unacknowledged and unprotested. In contrast with the tumultuous debate fostering the association's inception, the end of the HRVA came as a fading away, rather than as a noticeable disruption. With no funds in the HRVA's bank account, John Wyndham paid out of his account for the association's dinner in 1876, and there appears to have been no HRVA meeting after that year. The association's original purpose, to spur on Hunter wine, had been achieved. A maturing and consolidation had occurred in Hunter winegrowing – as hoped by the imaginers – brought to fruition by the experimenters.

> In contrast with the tumultuous debate fostering the association's inception, the end of the HRVA came as a fading away, rather than as a noticeable disruption.

Following the death of O'Kelly, the loss of the Northumberland Hotel as the venue for the meetings cannot be overlooked as a break in tradition for the HRVA, a group formed at the height of hopes signified in the patronage of grand edifices to enterprise such as the Northumberland.

The Northumberland Hotel was built on the north side of Maitland's High Street by Henry Sempill and licensed in 1843 to George Yeomans. The building is described as constructed from brick and weatherboard in the style of English inns, with a flagged courtyard overlooked by balconies, containing 20 bedrooms, a large stable block and a blacksmith's shop.[55] As well as the HRVA, the dining room of the Northumberland hosted meetings of the Hunter River Agricultural Society and visits from governors Fitzroy, Denison and Young, in addition to the formation of the Hunter River Steam Navigation Company, and theatrical performances. In 1874, the New South Wales government purchased the Northumberland Hotel to replace the Maitland Courthouse and gaol, as the original buildings were too close to the banks of the Hunter River and continually flooded. The long-ago demolition of the Northumberland means it is not possible to go to the site of meetings of the HRVA. Or to visit the cellar that for a generation held a concentrated array of Hunter wines, as Hunter wine producers developed their identity and their wines gained an international reputation. The former site of the hotel is west of the present Maitland Courthouse.

A changing vine and wine landscape

AT PATERSON IN 1872, THE GRAPES WERE EXCEEDINGLY LARGE AT harvest.[56] But there were other changes in the air and the larger concentration of vineyards was moving westward from the earlier sites of winegrowing. Many of the former large estates in the Paterson district were sold at high prices; for instance, Cory's Gostwyck in 1874.[57] Land reforms from the 1860s encouraged smallhold beef cattle and dairy farms, orcharding and viticulture. The northern rail line connected smaller farms away from the main rivers and roads (especially Rothbury and Pokolbin – at the Branxton and Allandale stations) with Newcastle. Grape production for fruit or wine became a more viable proposition as land selections of 40 to 60 acres were available on long-term lease-to-buy arrangements. Small landowners were their own labourers – rather than the earlier model of estates requiring a large landless labouring class – drawing on some seasonal labour but otherwise working at a manageable scale of economy, and often connected with family and social networks.

In 1876, the *Maitland Mercury* reported that although the dry season had been unfavourable for stockowners and ordinary agriculturalists, conditions were exceedingly favourable to the growers of wine.[58] Grapes were shrivelled from the heat, promising high sugars, which would be confirmed by use of Keene's saccharometer. This vintage promised to be remarkable for the extremely high quality of the wines – as propitious as 1858, a year revered among the oldest growers. At Oakdale and Côte d'Or, Malbec musts were 26 to 28 per cent sugar mass, or Brix.

At Dalwood's 70 acre vineyard, the 1876 harvest had so far produced 31 000 gallons of fine new wine, of outstanding 'flavour, richness and fragrance', with Hermitage 'and other fine sorts' at 25 to 28 per cent sugars.⁵⁹ 'The vintage at these vineyards during "picking days" presents a pretty sight, and when once seen is not easily forgotten', mused the *Maitland Mercury*. 'Boys and girls, and women, flock in by troops from the neighbouring farms, early in the morning, their pleasant faces give ample evidence that the work they are called upon to perform is both pleasant and profitable to them', continued the reporter, who may not in fact have ever spent a day picking grapes. Having said that, some joy and pecuniary benefit seems likely for the workers, as many as 100 of whom worked at Dalwood at the height of the picking season. As the *Maitland Mercury* reporter remarked too, one visitor had declared the sight of the vintage 'worth

Dalwood's 80 acres of vineyard stretched to the horizon of the Hunter river flats near Branxton. H.G. Ballard, Photographs of the Dalwood Vineyards near Branxton, New South Wales, Australia, 1886, State Library of NSW, Call no. PXD 740/8.

going miles to see'!⁶⁰ Apart from the appeal of the scene, the concentration of labourers – women and children from within the local community – emphasises the reach of the industry into the wider community of stockholders and other farmers thereabouts.

The aim of the founders of the HRVA in 1847, 'to gather every reputable and intelligent vinegrower in the district into friendly discourse', gave vital impetus to visions for a wine industry as opportunities arose to form wine businesses.⁶¹ The instigators of the HRVA, King and Carmichael, brought a certain élan to the early organisation, and King would surely have done more if he had not died suddenly in Scotland in 1857, during a return visit. Across the generation of the experimenters, men such as skilled horticulturalist William Kelman continued to advance the reputation of the region as a grape and wine producer, while Henry Lindeman laid the foundation for one of Australia's most famous wine brands. Members of the HRVA exchanged technical and cultural know-how that created an industry. Beyond this, they formalised the association of the Hunter region with 'wine' and, more broadly still, changed the emerging society of the region through sponsorship of German immigrants with skills to benefit the winegrowing community.

Chapter 5

A common people's paradise: German immigration from 1849

AT GLENDON, IN EARLY SPRING 1855, SARANNA SCOTT TOOK UP her pen to write to her husband Helenus, who was stationed for a year at Carcoar in the central western districts of the colony as a collector of Crown taxes. In reporting on the estate and household, Saranna described an enthusiastic laughter and chattering in German spilling from the kitchen, following the arrival that day of Peter and Regina Habig. ('Habbisch, as far as I can make out', wrote Saranna.) Peter 'seems to have been a labourer and to understand a vineyard', said Saranna, and the newcomers were engaged in lively conversation with 'John and Margaret' (whose German names she anglicised). Saranna supposed the couples were 'talking over the Varterland [Fatherland] together'.[1]

Peter and Regina, John and Margaret were among some hundreds of emigrants from south-west Germany who arrived in New South Wales from 1849, encouraged as agricultural workers through government bounties and settler sponsorship. These people were intended to fill the gap in the labour market created by the end of the convict labour system, and they, in turn, were promised 'Schlaraffenland, a common people's paradise, where everyone willing to work will be richly rewarded'.[2] (Given the racial prejudices of the British world of the period, Germans – rather than Chinese or South Sea Islanders – were preferred as replacements for convict labourers.)

While not all of the German men who immigrated in a subsidy scheme favouring married vinedressers did actually have this skill, or actually went on to work in Hunter vineyards and wineries, many did. These men brought their understanding of *Vitis vinifera* and winemaking to bear in a new climate, adapted their practices to colonial technologies, and (as did their families) negotiated a new language and culture. Others immigrated without government subsidies, and joined the labour market on arrival; while some who were sponsored to immigrate to New South Wales completed work contracts in other districts, and then relocated to the Hunter. Towards the end of the 19th century, a handful of immigrants who had left Germany mid-century were prominent members of the Hunter winegrowing community, most notably Anton Bambach, Carl Brecht and Richard Weismantel.

In the early years of the generation of experimenters in Hunter winegrowing, settlers such as Saranna and Helenus Scott were masters of servants (convicts, emancipists, British, German – men and women), but this changed from the 1860s. With the democratisation of land access, colonists of humble means were able to purchase small farm acreages on low-interest terms. While many colonial land selections were later abandoned – and there were German-born selectors among the insolvencies of the era – some former servants were, by the end of the generation of experimenters in the late 1870s, their own masters, and in some cases their children carried on their enterprises. Those German-Australians known to have carried on family businesses were Will Brecht and Andrew Weismantel.

Solving a labour crisis

It became clear in the 1830s that New South Wales authorities needed to intervene with subsidies for the immigration of labourers, and thereby spur economic growth through production from the estates of settler capitalists. When Governor Richard Bourke announced a bounty scheme for working class immigration in 1836, both the Macarthur family and John Dunmore Lang petitioned for subsidies to cover vinedressers and wine coopers from Europe. Although Bourke gave his assent for Europeans with winegrowing skills to be included in the scheme, colonial officials in London were very resistant to extending government funding to non-British immigrants. This was a problem for New South Wales vinegrowers, struggling in the absence of experienced vinedressers who could be obtained only from European wine regions. In an impassioned entreaty to the Colonial Office in Britain, Edward Macarthur – brother of William and based in London – pointed out that his family and some other colonists had invested heavily in winegrowing with the encouragement of government authorities. On these grounds the Colonial Office should not deny winegrowers financial assistance for skilled labourers, although these labourers would need to be recruited from winegrowing countries in Europe. With this argument, Macarthur succeeded in securing a reluctant extension of the bounty to a 'limited number of foreigners'.[3] Diplomatic relations between Britain and Germany fostered the choice of Germans, rather than Europeans from other wine countries. The first subsidised German immigrants in New South Wales were therefore six vinedressers and their families, engaged in 1837 by Edward Macarthur, and arriving at the Macarthurs' Camden Park property southwest of Sydney in 1838. Even so, this left Hunter winegrowers still bereft of skilled labourers and lacking the means to conscript them.

> In the 1840s, the New South Wales government – with a third of the colonial legislature engaged in winegrowing – began to relax policies on 'foreigners'.

In the 1840s, the New South Wales government – with a third of the colonial legislature engaged in winegrowing – began to further relax policies on 'foreigners', due to the success of the Camden vinedressers experiment and in the face of continuing labour deficiencies. In 1847, the New South Wales agent for immigration approved a proposal from Frankfurt-born Wilhelm Kirchner to recruit government-assisted vinedressers on the basis of a Certificate of Character. Kirchner's immigration prospectus aimed to stimulate the interest of German agricultural labourers, tradesmen, semi-professionals and professionals – although doctors, solicitors, scientists and clergy were warned that their prospects were not as assured as men of lower ranks of expertise.

Before his appointment to this recruiting role, Kirchner appears to have been employed at Glendon, and through this connection with the Scott family he had met and married the stepdaughter of Helenus' brother Walker Scott. Some historians have Kirchner and his wife Frances living at Ash Island in the late 1840s.[4] Another historian disputes that the couple lived at this Newcastle property.[5] Either way, Ludwig Leichhardt corresponded with Kirchner during his visit to Glendon in the early 1840s, and seems likely to have introduced Kirchner to the Scotts. Certainly, ties between Kirchner and the Scott family benefited the wider Hunter community.

In 1847, Kirchner visited Paterson, Singleton and Maitland – and other parts of New South Wales – to meet settlers seeking German labourers, particularly vinedressers, under the bounty scheme.[6] New South Wales settlers signed on to sponsor 95 immigrants. In the Hunter, Richard and Maria Windeyer requested seven vinedressers and a cooper; Henry Carmichael, three vinedressers; James King, two vinedressers and a cooper; William Caswell, a vinedresser; William Ogilvie, two vinedressers and a cooper; William Charles Wentworth, three vinedressers and a cooper for his Hunter property of Windermere; George Blaxland (brother of Gregory Blaxland, with property on Wollombi Brook and possibly in the Upper Hunter), two vinedressers; Walker Scott, five vinedressers; Archibald Windeyer at Kinross, one vinedresser and one cooper; Edwin Hickey, one vinedresser and one cooper; and John Taylor, four vinedressers and one cooper.[7] The requests totalled 31 vineyard workers and six men trained in barrel-making. There were two subsequent waves of immigrants fostered by Kirchner up to 1855. As the Scott family's import of vinedressers from 1849 to 1855 exceeded that of others in the Hunter region, it may be that they forwarded many of the German immigrants they originally sponsored to other settlers, although the circumstances of this are not recorded.[8]

In 1848, Kirchner, Frances and their Australian-born children travelled to his former home in Frankfurt – located on the Main, a branch of the Rhine River – staying with Kirchner's mother while he and his subagents visited rural towns to recruit immigrants.[9] People of this region were

Morpeth Wharf was a crucial transport hub. Edward Charles Close, Sketchbook of scenes of Sydney, Broken Bay, Newcastle and region, New South Wales, 1817–40 / PIC Drawer 8631 #R7249-R7276, National Library of Australia.

known as Hessens, for the wider state of Hesse containing Frankfurt and the winegrowing districts. The Hessens of the Rhein–Main were experiencing famine and the violence of a democracy movement fighting to remove an aristocratic regime; immigration to New South Wales seemed an attractive alternative to immediate turmoil, and to the uncertain future of labouring for wages on large estates.

Kirchner's prospectus advised Germans that Maitland was the principal town of the Hunter River district 'with about 3400 inhabitants and is an important and busy trading and production centre. In the neighbourhood of this town, on the Hunter and its tributaries is grown the best wine in the colony'. The prospectus included a favourable account of winegrowing in New South Wales, translated from a London newspaper. The newspaper report described wines for sale in the colony as either European or Cape imports, light bodied and consumed mainly by the working classes at a cost of 5 shillings per bottle. 'Now a great deal of trouble is being taken to bring to the colony people who understand the growing and making of wine', the prospectus explained. Vinedressers were offered contracts with colonial estate owners for two years for payment of £15 per year, 'a house to live in', and a weekly ration of '10 pounds of meat, 10 pounds of flour, 2 pounds of sugar, a quarter pound of tea and half a pound of coffee'. Coopers were offered wages of £20 per year. Women and children were welcome. (Parents arriving between 1849 and 1853 were charged extra for the passage of children aged under 14; whereas, in 1855–56, the New South Wales government paid for the passage of children up to this age, and these costs were then extracted from employers.)[10]

Interested Germans applied to Kirchner by completing a Certificate of Character (*Zeugnis*), written in German on one side and English on the other, signed by their Bürgermeister and church minister. The *Zeugnis* showed names, ages, a certification of health, and a declaration from the immigrant stating their colonial employer's name, along with documents from a consular official recording their departure from their home state.

Once Kirchner had attracted sufficient emigrants from the German states in 1848–49, he contracted ships to take them from London to Sydney. Four are known to have sailed with government-assisted Germans on board.

These people travelled by boat down the Rhine River to Rotterdam, Holland, and then by ship to London. The first of these vessels, the *Beulah* and the *Parland*, carried many couples and families who were destined for Hunter employers. In mid-April 1849, 'a number' of the immigrants from the *Beulah* were reported as arriving at Morpeth via riverboat up the Hunter, and 'many were immediately forwarded to their respective employers'.[11]

Kirchner and his family returned to Australia in 1851. In 1852 and 1853, he arranged for more immigrants through agents, and then sailed again to Germany for a very large recruitment drive in 1854 and 1855, making the return passage from this venture in 1858.[12]

In 1849 and 1850, Kirchner's immigrants included only a small number of private contract men for positions other than vinedresser. But from 1852 onwards, many more men from other trades gained passage to the Australian colonies; some with work contracts negotiated privately with employers, others speculatively. Those Germans with private contracts to work as farm labourers are less visible in the public record, although they are known to have been sent to employers throughout New South Wales, no doubt including the Hunter.

Sudwallisen *(New South Walers)*

FROM 1849, MORE THAN 160 GERMAN MEN AND THEIR WIVES AND children arrived in the Hunter through Kirchner's auspices. Historian Jenny Paterson has comprehensively traced the life-course histories of these German-born immigrants. They came from the rural working classes of the Rheingau (the winegrowing region on the Rhine River, at that time in the Duchy of Nassau, a south-west German state), the Grand Duchy of Baden (encompassing parts of north-west Mannheim-Heidelberg and north-east Weinheim), and from the portion of the Kingdom of Württemberg that extended along the Neckar River from Heilbronn south to Stuttgart.[13] By the 1870s, a democratic German nation emerged from the previous duchies, grand duchies, kingdoms and electorates left behind by the many Germans who immigrated to settler societies such as New South Wales. Rheingau people were primarily

Catholic and educated, but they faced poor prospects for land ownership in a changing Germany. Historical events in Germany and Australia since the 1840s make it difficult to untangle where in Germany the Hunter Germans came from, their given names, who they worked for in New South Wales and whether they became farmers, especially winegrowers. Identifying how immigrant Germans spelled their names, referred to their homeland and expressed thoughts about their new homes is intended to more clearly trace the connections between German immigrants' country of birth and their new lives in the Hunter.

One of the 'forty-niners', Josef Horadam, wrote to his brothers and friends about his journey to the Hunter. After arriving in Sydney in early April, he and his fellow passengers remained aboard their immigration ship until their sponsors arrived to meet them.[14] 'Then we travelled one night and one and a half days on a steamship' to Newcastle. From there, Horadam continued, 'on a tributary, then for four hours overland and so we arrived on our farm. It is a big farm, so big it takes a person one hour to ride around it'. The Hunter appeared, he thought, a 'most beautiful region'. Here, Horadam told his readers, Johann Wenz from Hattenheim already had a 28 acre vineyard (presumably this vineyard was owned by John Taylor, who sponsored Horadam's immigration, but Wenz had been naturalised in 1849 and may have purchased his own land). Wenz had arrived 11 years earlier among the first *Sudwallisen* recruited for the Macarthur vineyard at Camden Park. Horadam was sponsored to immigrate by Taylor at Lochinvar, between Maitland and Singleton, and this is where he wrote his letter.

As Horadam went on to explain:

The best thing in this country is that you do not have to terrace the land first, you just plough it and put the plants in. They don't dig like in Germany, they plough. We only have to clean the vines. That is the work the vineyard workers have to do, they only have to cut the vines by hand. We haven't got to cut trees down, they burn them out. The whole work is not more arduous than going for a walk. As they heard that I was a gardener, I was sent immediately to the garden. This is a most beautiful garden. It contains about 6 morgen

[3½ acres], a part of it is vineyard. I live at the entrance of the garden. This is a beautiful house built of bricks, it looks like a Swiss house. It has five rooms, a kitchen, everything furnished very nicely.¹⁵

Horadam's effusive account may have been solicited by Kirchner to boost further immigration, but it still affords a fascinating view of this new wave of immigrants joining the Hunter winegrowing community as well as some sense of how their work proceeded in the region's vineyards. Note too that Horadam referred to his skills as gardening, as distinct from vinedressing.

Gerstäcker visits Irrawang

IN 1851, THE FAMOUS GERMAN TRAVEL WRITER FRIEDRICH GERSTÄCKER visited the Hunter, during his tour of New South Wales and South Australia. Travelling from Newcastle by river, Gerstäcker passed through swampy country with low shrubs and, even though there were flowers, he found the vista to be 'sad and desolate'. Further upriver, the landscape took on a more pleasant character. Every now and then, small country houses emerged from the scrub, and 'the further we sailed upstream, the more cultivated I found the land; the withered trees left standing in the fields, the forest behind and the low banks all made the whole area look uncommonly like the Mississippi, though of course on a much smaller scale'.¹⁶

Raymond Terrace appeared to Gerstäcker as 'a flourishing little town'. He walked from there to Irrawang, as recommended by Kirchner and another German friend, to find King, who by this time had devoted some of his land entirely to viticulture and leased the remainder to tenants. From King's wine cellar, Gerstäcker tasted 'Irrawang 47' (a wine made from grapes harvested in 1947), a sort of fiery *Hochheimer*, and a red wine 'absolutely up to the standard of the *Aßmannshäuser*'.

Hochheimer – made from Riesling, and perhaps blended with some other lesser known white grape varieties – had long been exported from the Main region to Britain, where British drinkers knew it as Hock (although apparently the British called all German white wines Hock).

In this case Gerstäcker was describing King's wine as tasting like a very fine Riesling wine, although the grapes were more likely to have been Semillon. *Aßmannshäuser* refers to the fine red wines of the region of that name, made from Pinot Noir (known in Germany as Spätburgunder). King's red wine was probably made from a Shiraz grape or blend of red grapes rather than from Pinot Noir. Gerstäcker paid King a great compliment in comparing his red and white wines to the finest available in Germany, when it is hard to imagine that they tasted much like German wines at all. To King's credit, however, he had been experimenting with vinegrowing, winemaking, wine tasting and wine cellaring since 1836, and he would have known which of his wines to share favourably with his influential visitor.

Touring Irrawang on horseback with King, Gerstäcker met two German families who:

> talked very favourably about the land and assured me that whoever wanted to work would get by and do very well by it. On the other hand, they did not particularly like the 'life in the bush'. Whoever in Germany was used to the noise, or even just the more social aspect of life in the major or minor cities, and was perhaps a little too emotionally attached to the local places of amusement, will always miss those things, no matter to which country he immigrates. It is for this very reason that his personal circumstances are so much better than in the old country, because workers are in short supply, and consequently there is a dearth of the social life that the presence of people provides. Life in the bush has a character of its own, and like every other life has to be learned at least to be understood, and whoever is not self-reliant will not be able to feel at home here. The immigrant only needs to look on it as preparation for a better future, for his own hard work attracts neighbours, and in time 'society' forms itself.[17]

Who did Gerstäcker meet on his ride with King who found 'the bush' so unappealing? Those at Irrawang in 1852 were Johann Henkel, Peter Krahn, Heinrich Schwarz, Johann Georg Foesch, Johann Georg Carl Ulrich,

'Irrawang. James, King Esq, May 14th, 1852'. Conrad Martens, Scenes in Sydney & New South Wales, 1836–63, State Library of NSW, Call no. PXC 296, frame 18.

possibly one other man whose name is not recorded – and their families. Four of these men were assigned before Gerstäcker's visit, although which is not clear. In 1855, new immigrants were based at Irrawang. Johann Denzel, Joseph Krämer and Friedrich Müller.[18] Each of these Germans were from different township districts. They had likely become acquainted only once installed at Irrawang, which was remote from major and minor towns for people travelling on foot, or horse, or horse-drawn cart – preventing the socialising familiar in the more densely populated Rhein–Main.

Glendon friendships

SOME MONTHS BEFORE PETER AND REGINA HABIG ARRIVED, 'NANCY' Idstein joined the staff at Glendon. On this occasion, Saranna described a 'clatter of German for some time', as John and Margaret welcomed Nancy. This new member of the household 'settled well' thought Saranna, 'and has written a long letter to her parents to tell them all about it'. Most evenings Saranna taught her children lessons in one of the rooms of the Glendon homestead, and she invited Nancy to sit and listen while she sewed, which helped the German woman to quickly learn a few English words.[19] Saranna's daughter Rose Scott, later a central figure in the New South Wales suffrage movement (with Maria Windeyer's daughter-in-law Mary

Windeyer), portrayed these lessons as readings of English literature, and other discussions of scholarship that drew on the great library at Glendon.[20] 'Nancy' may have been Christina Idstein, the wife of an immigrant worker sponsored by Walker Scott, or more likely Christina's daughter.

Saranna and Helenus Scott relied on John (whose German name is not recorded by Saranna) for many tasks apart from vineyard labour. In 1855, John plastered and whitewashed the sitting room, repaired the dairy and floored a back room. He assisted with hitching the bullocks for ploughing, and then pruned grapevines. One day as John worked in the vineyard, Saranna wrote, her son Robert brought up to the house two ripe bunches of grapes. John expressed surprise that Glendon's grapes often fruited twice (which is common in tropical regions but impossible in Germany's cooler climate). He also found wine in New South Wales to be surprisingly expensive (at about a shilling per bottle) compared with a penny per bottle for ordinary wine in Germany. 'It is the dearness of everything that quite startles them', Saranna wrote in a letter to Helenus.[21]

> The Scotts' German worker John expressed surprise that Glendon's grapes often fruited twice (which is common in tropical regions but impossible in Germany's cooler climate).

In June 1855, John ploughed at Glendon with six bullocks (some borrowed from neighbours) – a process that mystified him at first. German ploughs were fitted with a wheel, so they could be operated by one man. Saranna reported to Helenus that John learned swiftly, and that he worked hard at ploughing, as the ground was dry from lack of rain and the soil had been compacted by cattle. John also slaughtered a heifer for the first time in his experience, and received a shoulder of the beast for his effort.[22]

Among the Aboriginal people who lived at Glendon was young Garry Owen (his age is not recorded) and, in June 1855, Garry and Robert Scott 'were lucky enough to find a bees' nest in the bark near the fence of the Vineyard paddock'. They brought home 'a little honey', Saranna relayed to Helenus, 'one bottle full; the comb was old and the honey dark coloured'. An Aboriginal man accompanying a visiting relative suggested the comb be left outside to allow more honey to drain out. John's wife Margaret thought it too much of a risk, though, that Garry and Robert might help themselves to their treasure, and she and John carried the comb to a food safe, away from mischief.[23]

When, in September 1855, Peter and Regina Habig arrived at Glendon, Saranna remarked that they were older than John and Margaret, and childless; Peter appeared heavy set and stood taller than John. Peter explained that John's home in the Duchy of Nassau was 'one half hour journey from my country'. Regina and Peter had travelled alone from the Hunter riverboat wharf (likely at Morpeth) to Glendon, and John had come upon them walking towards Glendon as he was riding by pony to Conrad Voelker's farm to borrow some washing tubs. Voelker had immigrated in 1852 as a cooper sponsored by Pike at Pickering estate, and by 1855 may have been tenant farming in the district.

Sadly, the beautifully woven fabric of life of Glendon that shimmers from the pages of Saranna's letters would be destroyed with her nervous breakdown in 1857, and the sale of the property for financial reasons in 1858 – after which the family relocated to Newcastle. Saranna's daughters were protective of their mother's sensitive disposition, with Rose Scott providing loving care for Saranna after the death of Helenus in 1879.

A German network among the estates

IN 1853, MICHAEL JESSER AND HIS WIFE MARIA JOINED THE FAMILIES OF two other German immigrants at Boydell's Camyr Allyn, those of Gottlieb Knödler and Joseph Frederich Eyb. All three families had arrived that year on the vessel *Johann Caesar*. According to Jesser, Boydell was a good master. Each German family lived in their own pinewood house and, after a 'good welcome' Jesser felt this new situation to be 'ten thousand times better than in Germany'. In New South Wales, he wrote in a letter to a friend, working class women did only domestic work, rather than also being required to perform field labour as in Germany.[24]

In 1855, Christian Carl Krust, based at Dunmore, also wrote to Germany, reminiscing about his long journey from Hamburg in 1853. He and his wife Johanna had sailed first to Sydney, and then via coastal steamer to Newcastle, followed by a further eight-hour passage upriver to Morpeth. At Morpeth, the pair was met by Johann Georg Schmid, who had emigrated in 1845 from Esslingen. Together they travelled by wagon to Dunmore – where Schmid had been employed for ten years. Like John at Glendon, Krust thought food, beer and wine were expensive in the Hunter, and wondered why wine was not sold from the wine press, as it was in Württemberg.

Krust expressed satisfaction that he had immigrated to a 'good area, and also to a good master'. Mrs Lang, wrote Krust, 'comes to see us often, but our master does so very rarely, as he leaves the vineyard entirely to Schmid, who – as I heard – grew him the best vineyard in the area right from the desert. Therefore our master does not worry about it'.[25] From Krust we also learn that in 1853 Johann Denzel and Wilhelm Wolfgang worked at Osterley, and 'were glad to be so near to us', and that Frederich Schnepple was at Tomago. Some time afterwards Schnepple relocated to Osterley, and Wolfgang to Tomago, although Krust does not explain why this exchange occurred.

Krust expressed affection for Aboriginal people who visited Dunmore. He liked these locals 'better than many Germans', but 'Schmid talks to them like children'. Krust found it sad that although Aboriginal man 'Tom King of Dunmore' was a 'king', he begged from settlers just as his 'subjects' did. Krust described Tom as polite when asking for tobacco or bread, wearing nothing but an old shirt, and as being unwilling to accept Schmid's invitation to sleep at the German man's house. Krust thought Aboriginal women were cleverer than their men, 'just as it is in Germany'. Schmid had explained to Krust that his house was built on the site of a former Aboriginal camp, but that he did not feel in any danger from Tom or the other local people. Jesser remarked that the 'native black people' were not Christians, and 'do not hurt anybody and have brought me a lot of honey'.[26]

While the record is too thin to draw many conclusions about how Germans and Aboriginal people interacted, the newcomers appear to have responded to the locals with curiosity and acceptance (and some infantilisation – in the case of Schmid at Dunmore), and a fraternity of sorts. It is harder to divine how to characterise Aboriginal responses to

Germans, but there appears to have been amity at Glendon, and at Dunmore and Camyr Allyn an expectation that Germans were no different from British settlers; interlopers with resources to share.

Peter Crebert, Newcastle

GARDENER PETER CREBERT (THEN NAMED GREBERT) ARRIVED IN THE Hunter among the forty-niners, at the age of 24, having emigrated with his wife Maria Louisa from Kiedrich in the Rheingau region of Nassau. After the completion of his immigration work contract, Crebert toiled in gardens and factories in and around Newcastle. He and his wife gradually purchased several allotments of land in the suburbs of Waratah and Mayfield centred on a neighbourhood called Germantown – creating vineyards and orchards to supply the port city's produce trade. By 1870, Crebert and his sons had constructed a winery and storage cellar; the family operated three wine presses, and the casks for the winery – sized from 300 gallons to hogsheads and smaller – were coopered on site.[27]

Crebert operated a wine shop in the front room of his house at North Waratah, from as early as 1872. In 1874, a representative of the Licenced Victuallers Association (chiefly publicans), prosecuted Crebert and a neighbour for operating their wine shops on a Sunday, in breach of the *Colonial Wines Act*. From evidence to the court (presided over by magistrate Helenus Scott, formerly of Glendon), we learn that the prosecutor visited Crebert's wine shop 'in the press-room at his house … and called for a bottle of wine'. Crebert inquired whether the customer would prefer light or dark wine, advising that 'the light was mostly drank'. The glass of wine cost 1 shilling and 6 pence, and Crebert accepted the invitation to share a drink with the visitor. Others were drinking at the premises, amid an array of wines available for sale. The vineyard for the wine shop (and a flower garden) lay at the back of the premises. Crebert pleaded not guilty to the charge but was fined £2, plus the costs of the court of 6 shillings and 6 pence, and 31 shillings for professional costs.[28]

There is some great irony that as wine came to be sold from the wine press – as Krust at Dunmore noted in the 1850s occurred in Germany

– this practice met with opposition from the competing business interests of publicans and clashed with the colonial history of moral restrictions on liquor trading on Sundays. While this is not to say Crebert ought to have broken the law, there appear to have been no lasting ill-effects as a result of his enterprise. He continued to be licensed to operate a wine shop – and was considered a respectable member of the Mayfield community – up until his death in 1892.[29]

The naming of Crebert Street, Mayfield, memorialises the German vineyard, winery and produce gardens of this working class district of the city.

Weismantels

THERE IS AN AREA AT STROUD, ON THE ROAD FROM RAYMOND TERRACE towards Gloucester, that used to be known as Johnson's Creek but was in the 1970s officially gazetted as Weismantels.[30] This is named for Richard and Catherine Weismantel, who immigrated to New South Wales in 1852, aged 34 and 30 respectively, from Erbach in Nassau. Richard and Catherine were accompanied on the voyage from Hamburg to Sydney on the vessel *San Francisco* by their children, five-year-old Valentin and three-year-old Katharina.[31]

Richard and Catherine Weismantel are named among German bounty emigrants assigned to William Kelman at Kirkton. Other Germans assigned to Kelman in 1852 and 1855 were Philipp Claudi, Jacob Dennewald, Christoph Krams, Friedrich Stecher, and their wives.[32] If Weismantel did indeed work with Kelman, this would have greatly benefited his skills as a winegrower in New South Wales, whether or not he possessed these skills in Germany. By 1863, the Weismantels were tenants on Australian Agricultural Company land at Johnson's Creek, although they were almost written out of this record. A company despatch that year to its directors expressed surprise that 'until recently' tenants made 'little attempt' at vineyards but now, happily, 'two have been commenced, one by Nicholls, *the other by a German*. Others will probably follow' (our emphasis).[33] Weismantel first made wine in 1865.[34]

In 1871, Richard Weismantel held a licence to distil brandy from grapes. This licence formed a crucial part of the business model for winegrowers such as Weismantel (and for many other Hunter settlers, including Alexander Munro), as poor-vintage-year grapes could be (quite respectably) converted into grape spirit, or brandy. Some brandy was added to colonial wines, but not to all colonial wines, as later temperance mores would have it. From 1874, Weismantel possessed a colonial wine licence – on the basis of the surnames listed, he was one of perhaps six Germans among the holders of 15 colonial wine licences issued that year.[35] Colonial wine shop licences were a business opportunity for Germans after their immigration contracts were completed, as much as for other colonists of humble means, as discussed in the previous chapter. Richard Weismantel's Hunter-born son Andrew continued to renew his family's colonial wine licence into the early 20th century, operating from the front room of the family home, which was located close to the road. Customers would tether their horses in a paddock opposite the house.[36]

Anton Bambach, Eelah (Maitland)

ANTON BAMBACH ARRIVED IN THE HUNTER IN 1849, AT 26 YEARS OF AGE. By 1865, he managed Eelah vineyard and winery near Maitland. A report in the *Maitland Mercury* (presumably by John O'Kelly) portrayed the property as flourishing under the care of Bambach: 'a vigneron of considerable experience'.[37] The grape varieties at Eelah were Black Hermitage 'ocrean' (perhaps Aucerot), Shepherd's Riesling and Lambrusquat.

The Eelah wine press was a closed room in the wine house, pictured by the *Mercury* reporter like so:

> a length of an ironbark tree has been hollowed out like a trough [and] into this a strong lid has been fitted, and the pressure necessary to extract the liquid from the grapes is obtained by a long layer of the same material as the trough, which is easily worked by a second and smaller lever.

This press appears to operate on a similar pressure principle to log presses, such as the one used by the Drayton family at the turn of the 20th century, but on a smaller scale. With Bambach's press, the grapes were passed through 'holes perforated near the bottom of the trough', to form the must to be fermented in barrels.

The operation at Eelah benefited from the investment of capital by owner JR Nowlan, as well as the expertise of Bambach. Nowlan had sent to England for the wire used in trellising the Eelah vineyard with a post and wire system of espaliering, which was at this time quite rare in the region. Although espaliering is the rule rather than the exception in contemporary vineyards, this new approach merited comment from the *Maitland Mercury* reporter, as producing more evenly ripened grapes than the post-only trellising of most other vineyards; and Bambach did, indeed, produce many award-winning wines over the next quarter-century.

Carl Brecht, Rosemount (Denman)

IN 1855, 47-YEAR-OLD JOHANN PETER CARL BRECHT, HIS WIFE CHARLOTTE and their three children arrived in Sydney from Württemberg as free immigrants. That is, although they immigrated at the crest of the mid-century wave of arrivals assisted by subsidies, they paid their own passage.[38] Brecht is believed to have worked for William Dangar at Turanville, Scone (not a known winegrower, so perhaps Brecht did not tend vines at this property). Brecht's 1862 certificate of naturalisation lists his residence as Muswellbrook.[39] (Only naturalised citizens could own land.) Brecht joined the Hunter River Vineyard Association in the later years of the organisation.

While little record remains of the purchase and naming of Brecht's Rosemount, a *Sydney Mail* reporter visited in 1891, by which time Carl Brecht had passed away, although his widow remained on the property, now managed by her son, Will. According to the story in the *Sydney Mail*, Rosemount wine not only reached a ready market in the north and northwest of the colony, but had won 350 awards in competitions, from local wine shows to international exhibitions. Brecht's vineyard of 36 acres was trellised with post and wire and contained 'hermitage, muscat, sherry, Madeira, Pineau, Shiraz, Shepherd's Riesling and Lambruscat'. The grapes were picked at natural sugar percentages of 20 to 33 per cent, with the Madeira and Muscat grapes achieving the highest gravity. Brecht added brandy to his port-style wine, and otherwise sold his products as pure wine, without the addition of distilled spirit.[40] Brecht also made a prickly pear brandy.

Rosemount's 1891 vintage produced 15 000 gallons of wine, at 1000 gallons per day, using clean, modern equipment rather than – Will Brecht sniffed somewhat disdainfully – 'old Scriptural machinery'. Rosemount wines were bottled at three years of age, and younger wines were kept in bulk.

Trading places

IT IS A MATTER OF FOLKLORE IN THE HUNTER THAT MANY OF THE MEN who immigrated under the scheme operated by Kirchner lied about being vinedressers, and that this reflected poorly on their character. There is, however, an explanation for the false statements made by some of the men. Only vineyard workers and coopers received subsidies in the New South Wales bounty schemes, and in turn subsidised labourers were required to be married. The subsidy for labourers in winegrowing, but not in other trades, could be the reason that some immigrants falsely stated their skills as vinedresser. Claiming to be a vineyard worker or cooper more readily gained couples and families a passage on ships to Australia. Perhaps the Bürgermeisters and ministers who signed untrue claims of emigrant skills in vineyard and winery trades were trying to help people seeking to make better lives abroad.[41]

There is a further possibility. As settlers in New South Wales sought general labourers as well as workers for vineyards and wineries – and the immigration scheme conditions were more favourable for vinedressers and coopers – there may have been some prompting from immigration agents or subagents for men to falsely state their trades. Settler capitalist families with known vineyards such as the Macarthurs and Scotts requested many more families than they required through the bounty schemes for German immigrants. Jenny Paterson is convinced that some immigrants were passed on to other settlers who did not have vineyards, as a solution to labour scarcity. Informal transfers of vinedressers appear to be common enough. For example, although Christoph Schneider was introduced to the colony by James and William Macarthur at Camden, there is evidence that this man's child was born at Thomas Holmes's Oakendale in the Hunter in 1853. The immigrants themselves may also have applied pressure in their desire to relocate to stay near relatives.

By 1855, changes to immigration schemes meant that any original, 'official', sponsoring landowner who allowed an unofficial transfer would have required the 'receiving' landowner to make a payment. This represented the amount of money owed to the government by the original sponsor over the two-year hiring period for each adult male worker in the immigrant's family (minus the government subsidy).[42] Without payment from the receiving landowner to the original sponsor, the sponsor would be out of pocket. No evidence of such payments between masters of estates has come to light, although settlers surely found means to secure the labourers they required for their large properties.

German immigration to the Hunter in the mid-19th century may have been a means of bolstering the strong desire among some colonists to develop a wine industry *and* a boon for settler capitalists facing a continuing labour shortage.

Paradise found

GERMANS WERE THE LARGEST DEMOGRAPHIC OF NON-BRITISH MIGRANTS to Australia in the 19th century; larger in number than the Chinese – the second largest group – and far greater in number than the Italians, who might have brought other wine traditions.[43] On arrival in the Hunter between 1849 and 1855, the Germans in our story were dispersed in pairs or small groups to isolated properties. They quickly learned English and otherwise changed their lives to fit their new circumstances.[44] Although many Hunter Germans may have maintained pride in their Germanness, they did not form German communities as visibly as some South Australian German-Australians, who transplanted established social hierarchies, re-created visible culture such as architecture, and carried on German language practices in religious worship and newspapers – reinforcing the idea of Germany, not Britain, as 'home'.

Differences in German winegrowing practices, and the warmer climate at latitudes of 33 degrees to 34 degrees south, compared with the Rheingau at 50 degrees north, led to disagreements in other wine regions of New South Wales between British masters and German servants over how to manage vines and make wine.[45] Yet, for the most part, Germans greatly assisted settlers in Hunter winegrowing, and in many other tasks on their estates. The Germans themselves formed friendships on these estates, and between properties. When their work contracts were completed, many became naturalised citizens in order to own land, and became respected members of the wider community.

> The Hunter winegrowing community benefited greatly from German labour, and then from German participation as vineyard owners and winemakers.

The Hunter winegrowing community benefited greatly from German labour, and then from German participation as vineyard owners and winemakers. John Dunmore Lang believed his brother Andrew 'would never have had a vineyard but for this valuable man', George Schmid. Lang's winegrowing efforts did not continue, however, and Schmid's naturalisation papers have him at Henry Carmichael's property, Porphyry Point, in 1858.[46] Historian Cameron Archer argues that in the Paterson district, the more lasting impact of German immigration came not from the existence of vineyards and wineries, which had largely faded from that area after the 1870s, but in the social contribution of German people, enriching district settler culture.[47] In other districts, there was a more enduring German mark on winegrowing as well as on communities. The eager young labourer at Dunmore befriending Aboriginal locals, the disgruntled men at Irrawang, and the couples laughing in the kitchen at Glendon – these people all brought deep understanding of life in an old wine country to the vineyards, wineries and wider social milieu of a new wine country.

Chapter 6

New worlds of wine, at home and abroad: first wave globalists, 1877–1900

IN 1878, THE *SYDNEY MORNING HERALD*'S BRITISH-BORN REPORTER John Stanley James visited Dalwood. Enigmatic, restive and by all accounts a tenacious and fearless journalist, James styled himself 'The Vagabond', reflecting his persona as a well-travelled maverick of the wider British world.[1] Yet for all of his wry impartiality (and a later, incisive anti-colonialism against French occupation of New Caledonia), James fell just as soundly under the spell of the scenic 'glories of the Hunter Valley' as others before him – passing there 'one of those pleasant day oases in my rugged vagabond career, which will be for ever impressive on my mind'.[2]

James encountered the Hunter as winegrowers found a strengthening market in the Australian colonies and began to join a new globalised wine trade. From the 1860s to the 1890s, the colonial Australian economy experienced a long boom; the 1860s also mark the start of the first wave of wine globalisation, as defined by economist Kym Anderson and economic historian James Simpson, which continued until World War I.[3] In the generation from 1877 to 1901, Hunter wine's colonial, intercolonial and international reputation soared. Hundreds of thousands of gallons of barrelled wine flowed outwards from the wineries of the region in every direction, some across the globe – while certificates, medals and trophies from wine competitions, as well as visitors, flowed in.

The first wave of wine globalisation

ECONOMICALLY, WINE GLOBALISATION MEANS A RISE IN THE PROPORTIONAL quantity of wine that each wine country exports, compared with domestic consumption. Until the 1860s, European wine was largely consumed close to its places of production, or no further afield than the borders of each wine-producing country. Only about 5 per cent of each wine country's production was exported each year. Then, in the 1860s, came the sudden and widespread destruction of European vineyards as a result of diseases and pests that will be explained below. This disturbance, on the one hand, devastated livelihoods in European winelands along the length of the wine value chain; that is, the forward flow of grapes produced by workers in vineyards for vineyard owners, the processing of these grapes by winemakers, and the flow of wine to traders and drinkers (along with suppliers of other products and services at each link in the chain). At the same time, the loss of grapes for traditional wines led to a major reorganisation of patterns of wine trade to replace them – creating the first wave of wine globalisation, as wine traders sought to maintain their end of the value chain by finding alternative sources of grapes and wine. Countries newly supplying wine to Europe saw their annual export levels rise to as high as 12 per cent. (By comparison, in the contemporary second wave of wine globalisation that began in the 1990s, two-fifths of the world's wines are consumed outside their country of origin, and Australia exports about two-thirds of its production.)[4]

The search for replacement wines to fill customer orders offered some new opportunities for the grape-growing areas of the British empire at the Cape in southern Africa and in the Australian colonies, including the Hunter. A few Hunter wines were sold in Britain and Europe as a result of personal connections, but it was a pull from the British wine trade, responding to new market demands, that led to an appreciable quantity of Hunter wines joining the globalising flows of wine trade in the northern hemisphere – rather than a push from the Hunter. Before the budding of British interest in the region's wines, Hunter winegrowers had sold their brands widely in Australia and through the Pacific shipping trade as fine styles through to ordinary.

In 1878, a half-century after the importation of *Vitis vinifera* from Sydney's Botanic Gardens to the Hunter, the vines at John Wyndham's Dalwood property alone were of an extent that would blanket the entire acreage of today's Botanic Gardens site on Sydney Harbour. A combination of vast vineyards and modest acreages patchworked the Hunter Valley. If we trace an imaginary line on a map of the Hunter at this time, joining the outermost locations in which vines were planted, across the entire region, we can see the shared influence of British and German winegrowers. The vineyard and winery of Peter Crebert in the Newcastle suburb of Mayfield sat at the southern, port-side limit of the region. Weismantels at Stroud marked the eastern extent of plantings. Brecht's Rosemount guarded the northern edge of the region. Bambach of Eelah, and many other Germans, owned and managed vineyards clustered heavily in the Lower Hunter. Continuing to trace the line anticlockwise, from Brecht's Rosemount near Muswellbrook, to join up with Crebert's backyard in Mayfield (curving outward in the south-west to encompass Broke, Pokolbin and Wollombi), this final sweep captures the predominance of British settler winegrowers in the areas of Pokolbin and Rothbury: the many members of the Wilkinson family, the Tyrrells at Ashmans, the Draytons at Bellevue, and so on. Within this circumference were the vineyards and wineries of the Lindeman family at Gresford near Paterson, and other large businesses such as Dalwood, Bebeah and the Green family's Allandale cellars.

As wine from the Hunter became increasingly global, local ties within the winegrowing community were also growing stronger and more numerous through marriage, friendship and business. At the cusp of the new generation of first wave globalists in the Hunter, The Vagabond came to sing the praises of Dalwood.

'I wish that I lived in a wine country'

IN 1878, JOHN STANLEY JAMES CAUGHT THE TRAIN FROM NEWCASTLE TO Greta. The journey along this line passed through the 'most fertile part of Australia' that he had yet seen. James was amused that the rail stations were crowded very closely, every 2 or 3 miles, but delighted that at 'every little station along the line we saw baskets of butter on the platform, waiting for the train to Newcastle ... In all the paddocks there were fine cattle and horses, but few sheep, which one took to be a good sign ... There is an air of substantial and long-settled prosperity around'.[5]

James rode by horse and buggy from the railway siding at Greta, 'through the bush' to 'the celebrated' Dalwood, with its old homestead perched on a 'goodly site on the rising ground above the flats' overlooking 'the open valley formed by a bend of the Hunter', amid low surrounding hills. On the other, eastern side of the Hunter were the many small cabins of farm workers, and on both sides of the river bank were 'great patches of lucerne and maize, growing as luxuriously as in Egypt'.[6] A notion of this stretch of the Hunter as an antipodean Nile Valley (but with small eucalypt-shrouded hills, tinged bluer with distance) was heightened by his host John Wyndham's accounts of severe flooding 50 years earlier, evoking for James the ancient creation of rich farmland soils through the seaward wash of alluvial deposits.

> On both sides of the Hunter River were 'great patches of lucerne and maize, growing as luxuriously as in Egypt'.

James portrayed Dalwood as the quintessence of settled colonial life, forged by an earlier generation of 'pioneers'. The wine was 'pure juice of the grape', the 'children are beautiful and healthy', the young ladies almost the most beautiful in the Australian colonies. Wyndham recalled his family's early settler experiences, when the cruelty of some masters turned convicts to further crime, so that the windows of Dalwood House were barricaded 'with loopholes for musketry, standing a siege against bushrangers'. Now, instead of holes for firearms, Wyndham's office walls were thick with certificates from agricultural societies and international exhibitions, prize cups and medals (most impressively a silver medal from the Paris *Exposition Universelle de 1867*, of which John Glennie of Orindinna was especially proud).

Despite the high summer heat at Dalwood, wrote James, 'Everything is so soothing ... there is more green here than I have yet seen in Australia: and the iced wine at my elbow is remarkably good'.[7] Tasting wine on a hot afternoon was a fine occupation for a 'seasoned vessel' such as James and, 'what between the white and the red wines, I put in the time very pleasantly ... I wish that I lived in a wine country'.

Aboriginal people and wine country

WHEN THE VAGABOND VISITED DALWOOD, THERE WERE IN ALL likelihood Aboriginal people among the working community, although James did not concern himself to find out. To James the region appeared as a settler oasis of restful peace compared with 50 years earlier. The transformations of 'Australian history', as The Vagabond surely understood them, had evidently moulded an idyllic agricultural landscape and then moved on. 'History', such as the frontier wars of settler occupancy – as well as the gold rushes, the boom in wool production, the growth of cities, the action of parliaments – happened elsewhere, away from the Hunter. Seeing the Hunter this way, however, shows that the force of the colonial frontier did *not* wash away Aboriginal people. Those who survived settler occupancy continued to live in the region.

At Camyr Allyn in 1882, James Boydell recorded that 15 Aboriginal people were collected by the manager of Maloga Aboriginal Mission, in Victoria, and moved there, far from their ancestral lands.[8] There are few other details of this event. The wider context is that in the late 19th century colonial governments increasingly funded church-run missions for religious instruction on lands reserved for Aboriginal people. The removal of the Camyr Allyn locals to Victoria before missions were established in the Hunter appears to have been at Boydell's request, in response to colonial policies controlling the lives of Aboriginal people. While there is no evidence of a connection between these two events, following on from the removal of the Camyr Allyn group the Aborigines Protection Board in 1883 began discussions to establish an Aboriginal mission at Karuah, Port Stephens.[9] Land on the Karuah River was gazetted as reserve lands for Aboriginal use in 1898.[10] In the early 20th century, St Clair Aboriginal Mission opened at Mount Olive, Singleton. Although Aboriginal men from Sackville Reach Mission in the Hawkesbury district picked grapes for Italian surgeon Thomas Fiaschi's Tizzana wine enterprise at Windsor from the 1890s, Hunter winegrowers do not appear to have employed people through contact with Aboriginal missions and Aboriginal people were denied freehold ownership of land.[11]

Forces of nature, disruptions to trade

THE CATALYSTS FOR THE 19TH CENTURY WAVE OF GLOBALISATION WERE the European infestations of powdery mildew, followed by phylloxera. Only a nuisance for colonial Australian grape producers from the 1860s, powdery mildew almost wiped out Portugal's much larger and older wine industry in Madeira in the 1850s, and proved perilous to production communities in parts of mainland Europe. From the 1850s, producers in France – starting in the southern Rhône region of Languedoc – also began to experience unexplained vine die-off. Unlike powdery mildew, which appears as a white substance on leaves and fruit, this second cause of vine damage was concealed within vine root systems, nearly invisible to the naked human eye, and remained elusive for many years. Once the cause of vine damage had been identified as the aphid-like grape phylloxera (*Daktulosphaira vitifoliae* or *Phylloxera vastatrix*), it became apparent how this tiny insect – uninvited and unforeseen – could remain undetected until vines began to inexplicably lose their vigour. Finding and executing a solution among vinegrowing communities deeply set in their ways proved as difficult as the initial identification of the problem.

Grape phylloxera is native to North America, where it feeds on other *Vitis* species that evolved in ecological balance with the insect. But once unleashed in Europe, as a result of the 19th century's enthusiasm for experiments in plant acclimatisation, phylloxera achieved widespread destruction of vast *V. vinifera* monocultures. In the 1860s, European plant scientists determined that the solution for phylloxera was prevention rather than cure. This involved the vine-by-vine grafting of *V. vinifera* cuttings onto rootstock of resistant North American *Vitis* species: *V. labrusca* and *V. rupestris*.

Phylloxera was first detected in Australia in 1877, near Geelong in Victoria. By the 1880s, the insect had been found in the southern vineyard districts of New South Wales, including Camden, south-west of Sydney, where William Macarthur and others had produced wine since the 1840s. Hunter winegrowers, for their part, were enraged at the New South Wales government's regime of regulated vineyard inspections and compulsory destruction of phylloxera-infected vineyards. Despite suggestions that phylloxera may have infested vines in the Hunter, the region remained officially phylloxera-free, with some ungrafted vines still in existence.[12] There are parts of France and other European wine countries that also escaped the spread of phylloxera.

The consequent grafting of much of the world's *V. vinifera* to *V. labrusca* or *V. rupestris* led to a rise in the cultivation of these species, and added to the cost of establishing vineyards. In New South Wales, the colonial government assisted the grape-growing industry from the late 1890s by funding the cultivation of phylloxera-resistant *Vitis* species and the grafting of *V. vinifera* cultivars on to this rootstock.

The breaking of wine ties to place

THE FIRST WAVE OF GLOBALISATION WAS A TIME OF ENORMOUS DISRUPTION in the wine world, which saw the development of new practices for applying sulphur to grapevines in order to combat powdery mildew, and the introduction of new *Vitis* species into the rootstock foundation of vineyards to combat phylloxera. Amid massive vineyard loss and replantings there were interruptions to wine trade traditions, old and new.

The search for new grapes and wines led to an upscaling of trade – from the traditional local and national level of circulation of grapes and wine in European wine countries, to a more complex criss-crossing of international shipping routes transporting wine from places of production to places of blending and then on to places of sale. This greatly distanced wine from production localities, erasing at each stage of transport the original identity of wines as tied at least to a vineyard region, but at the same time meeting the demands of wine customers. Most wines traded within the British world before the 1860s were known for their vineyard regions, if not for particular fine wine estates. Even the vaguest of connections to wine places in wine names were loosened, and in some cases broken. When export of even ordinary wines from mainland France slowed as a result of vine die-off, much of the gap in supplies was filled by products from France's North African colonies, such as Algeria. This new practice is relevant to Hunter wine history in that it disrupted the international reputation of Europe's wines built on grape origin from *regions*. The substitution of wines from France's African colonies led to some terrific contortions of logic about French regional and national wine identity.[13] Some wine merchants simply concealed the geographic origin of wines from North Africa, and elsewhere.

Once French wine production recovered from devastation by pests and diseases at the turn of the 20th century, grape producers in traditional wine regions began the long task of restoring the reputation of their grapes for the styles of wines that could be made from them. This was achieved in 1935 with the creation of the *Appellation d'Origine Contrôlée*, or AOC. The AOC defines regional boundaries and methods of vine cultivation as a form of intellectual property, to protect against the fraud that raged during the chaos of the late 19th century. (The now international system by which fare from particular places is defined on the basis of geography and husbandry is called geographical indications.)[14]

As the deep ties of tradition that bound vineyard workers and owners to winemakers and merchants (and drinkers) in Europe were tragically unravelling from the 1860s, such ties were strengthening in the Hunter. In the same period, the region's wines were becoming refined into more nuanced styles and flavours, and gaining a reputation as desirable among Australia's metropolitan sophisticates.

The Bordeaux Exhibition of 1882

BY THE EARLY 1880S, MELBOURNE HAD GROWN INTO ONE OF THE LARGEST and most vibrant cities of the British world. Visiting there, English-born journalist Richard Twopeny saw that 'a good deal of sherry and port – even more brandied than for the English market – is drunk' and observed that socialites were simply obsessed with Champagne. Searching out the better wines from the ordinary, Twopeny found that 'white hermitage from the Hunter River district in New South Wales, at 15s. a dozen, is also as good as one can wish, short of a *grand vin*' (or fine wine).[15] France's greatest fine wines were Champagne (from Champagne), Bordeaux (from the extensive Bordeaux wine districts) and Burgundy (from the Bourgogne).

It mattered a great deal to Hunter winegrowers that the annual International Exhibition of 1882 took place in Bordeaux. This river-port city serviced an influential hinterland of many high-quality vineyard estates and wine brands that lay at the epicentre of British wine export from France. In 1855, France's Emperor Napoleon III requested a classification of France's fine wines to guide British wine trade. The tiered list of *cru* (which is translated into English as 'growths'), from first to fifth level of named wine estates, was unveiled at the 1855 Paris Exhibition. In this hierarchy, first growth wines (*grand* or *premier cru*) were the most prized and luxurious. The 1855 Bordeaux Classification has largely framed international notions of fine wine since that time.

> Wines made by James King at Irrawang debuted on the continental stage alongside William Macarthur's Camden wines at the 1855 Paris Exhibition.

Wines made by James King at Irrawang debuted on the continental stage alongside William Macarthur's Camden wines at the 1855 Paris Exhibition, with each wine receiving scores of below 12 out of 20, excluding them from awards.[16] By contrast, in Bordeaux in 1882, there were 32 New South Wales wine exhibitors, and many received prizes. These prizes ranged from honourable mentions to gold medals (a gold medal scores above 18.5 out of 20, a silver above 17 out of 20, and a bronze above 15.5 out of 20).

The Hunter's gold medal winners at Bordeaux in 1882 were Carl Brecht's Rosemount, for an 1880 Red Hermitage; James Kelman from Kirkton, for this second generation wine family's 1876 (red) Hermitage; Alexander Munro's Bebeah (with an astonishing 31 wines on show), for four new red wines – Hermitage, Lambruscat, Malbec and Petit Verdot; Philobert Terrier, for an 1875 (red) Hermitage; and John Wyndham, for a selection of three Petit Verdots from Dalwood, of vintages 1881 and 1882.

Silver medals were awarded to François Bouffier at Marcobrunner, Theophilus Cooper at Oswald, James Doyle at Kaludah, John Hill at Hannahton and John Wilkinson at Coolalta. The Lindeman family entered wines from their vineyards at Cawarra and at Albury, and it is not clear whether their silver award pertained to Hunter wine. Those receiving bronze awards were Henry Carmichael's sons at Porphyry, AE Davies at Mount Huntley, and Lionel Stephen at Ivanhoe. Some entries went unawarded.

The cost of vineyard cultivation varied greatly. At Bebeah, Munro's manager William McKenzie kept costs to the average of £7 per acre, while Wyndham's costs were between £5 and £20 per acre, across the range of wine styles he produced. Hunter wines were sold chiefly in barrels, by the gallon. (Bottles and corks were still difficult to obtain, and Wyndham's records indicate the painstaking process of establishing regular supplies of these necessaries for a bottled trade.) Carmichael's and Brecht's wines received 10 shillings per gallon – the highest price, along with the wines of John Hill. Wine grape yields per acre, recorded for the Bordeaux Exhibition entries, ranged from 800 gallons per acre at Dalwood, to 166 gallons per acre at Oswald.

Henry Bonnard, executive secretary of the Bordeaux Exhibition committee for New South Wales, knew there would be considerable scrutiny of Australian wines, and no small degree of incredulity from French judges at the existence and taste of wines from a British colony. Yet Bonnard found that the entries were treated with respect by experts and the general public. The three Bordelais judges awarded each of the gold medals for New South Wales to Hunter reds, and made some remarks about the quality of the wines overall. They were surprised that Hunter wines were lighter in body than Albury wines, and presumed that this was owing to the region's maritime environment. The judges recommended that the fruit for the red wines be fermented with stalks, to provide more tannins – which would structure the wines for ageability. They noted that the wines did not lack body (or viscosity) from the weight of alcohol. To reduce a flavour of eucalyptus in the wines, the judges suggested more frequent ploughing of vineyard soils. Ultimately, the Hunter wines were described as 'really perfect ordinary table wines'.[17] Not the *grand vin* of Twopeny's more expansive remarks but, still, a sound vote of confidence from judges at the centre of the international wine world, when they might instead have been dismissive, or worse. Bonnard noted that the visiting public at the Bordeaux Exhibition were curious about New South Wales wines and spoke favourably of them.

Bonnard also delivered wines to the Amsterdam International Exhibition in 1883, at which three Hunter producers were represented: George Campbell at Daisy Hill and the Holmes family at Wilderness (both of Rothbury), and Montague Parnell at Maitland. Remarkably, at this time the 20-year-old Holmes vineyards were 76 acres in extent. Parnell cultivated 25 acres of vines, and Campbell a modest 10 acres.

Bonnard reassured the members of the New South Wales Legislative Assembly – to whom he delivered his 1884 report on the Bordeaux Exhibition – that the French wine jury considered Australia's transition to a wine-producing country to have been the most rapid, and most successful, of any nation undertaking this conversion in the global age of wine. The report advised the colonial government how to proceed with policy on the wine industry at a time when New South Wales wine production had almost halved because of drought and phylloxera – although the colony did not receive the formal training college for viticulture and winemaking suggested by Bonnard. (In 1883 just such an institution – Roseworthy College – had been established in South Australia.)

From 1890, the Department of Mines and Agriculture did however give considerable attention to the matter of vines and wine. These measures included the appointment of a colonial viticulturalist, first Frenchman J Adrien Despeissis, until 1895, and then Italian Michele Blunno, from 1896. Despeissis had studied at England's Royal Agricultural College and at Louis Pasteur's laboratory in France. Blunno had studied viticulture in France, and focused on eradicating phylloxera and advancing the adoption of technology in colonial winemaking.

Nicholson and the eyes of the world

IN THE LATE 1880s, BRITISH WINE MERCHANT PB BURGOYNE & CO opened an office in Adelaide. By this stage, South Australia – which was not infested with phylloxera – was emerging as Australia's biggest wine-producing colony. Large South Australian family wine firms such as Penfold's and Thomas Hardy & Sons had emerged in the same era as the Lindeman's wine company in the Hunter. Much of an upwards trend in wine production for the whole of Australia was due to South Australian output. Burgoyne sent Arthur Nicholson from London as the export agent for Australian wines in 1888. From that year, Nicholson travelled throughout South Australia and Victoria, and finally to the Hunter, in search of wines suitable for export. Burgoyne's would become the principal engineer of the new rivers of wine flowing from Australia and South Africa to the British Isles.[18]

Nicholson arrived in the Hunter in 1889, and left brief notes with key points about 19 Hunter properties. Many of the wineries he visited are well known from the stories we have told so far (sweeping from the Upper Hunter to the Williams River district): Rosemount, Kirkton, Kaludah, Wilderness, Bebeah, Dalwood, Eelah, Porphyry and Kinross.[19] Some others are less well known.

At Rosemount, found Nicholson, Brecht made good wine, all of which had a market – and there was none for sale.

Nicholson noted that the Kirkton vineyard was the oldest in the Hunter, owned by James Kelman but managed for some years by William Wilkinson, and planted on very sandy soil. Some vines were trellised and others remained as an older 'gooseberry style'; pruned as a standard shape, with a central trunk. The varieties were those William Kelman preferred in the 1840s: Red and White Hermitage, as well as Shepherd's Riesling (Semillon) and Verdelho. Yields were at least 450 gallons of wine per acre of grapes, and up to 700 gallons for the Semillon. Kirkton's Blanquette had been known to yield 1100 gallons per acre. Those wines noted as excellent were an 1886 Riesling, and Red Hermitage from 1885 and 1886. An 1887 wine was 'watery but clean'. Kirkton wines were sold under contract to a company called Stanley & Co (Littlewood) (this appears to be an American-based firm of general merchants) subcontracted with Hutchens & Co at 84 Pitt St Sydney, and at 3 shillings per gallon. Even from these brief notes of Nicholson's there is an impression of the connections forged in the wine value chain. The Kelman family did not have a Newcastle or Sydney agent, unlike John Wyndham and the Lindeman family, and negotiated instead to sell wines through general traders tied to wider business networks.

James Doyle's Kaludah vines were staked and trellised on a slope, on soils of part-chocolate (presumably brown basalt) soil, part-sand, intermixed with limestone and gravel. Most of the grape varieties were Hermitage, along with some White Shiraz (Trebbiano) and 'Vedentro'. The cellars were hot, but the casks good and clean; the wines were saleable, but Doyle did not want to sell. A grower called Onus at Lochinvar grew 10 to 12 acres of vines. His wines were good, but he had only the 1888 vintage left for sale.

The Wilderness vineyard of the Holmes family at Allandale was planted on a slope of sandy chocolate and limestone, with Hermitage (Marsanne or Roussanne – this is not clear), Shiraz and Verdelho, averaging 600 gallons per acre. The wines were sweet, although no spirit was added.

Twenty-three years after the first vintage at Bebeah, William McKenzie continued to manage Alexander Munro's vast estate, now expanded to 80 acres. The vines were planted on a slope of friable black soil, trellised with stakes and wire. Nicholson felt the cellar was well built but too hot. Bebeah's casks varied from 500 gallons to one that held 3500 gallons. The nature of the notes suggests the casks were coopered from cedar and beech – which raises intriguing questions about the taste of the wines: Hermitage, White Shiraz, 'Ved.' (Verdelho) and Pinot, and 'a little White Muscat of Alexandria'. The 1886 tasted good, fruity, but the 1887 was weak and watery. A white wine tasted 'somewhat coarse', a red 'excellent'. The 'Ved.', wrote Nicholson, had 'a curious bouquet', but was 'good quality'.

Moving on to Dalwood (with a new manager since John Wyndham's death in 1887), Nicholson's remarks are more extensive than for other sites. He noted the vineyard as 80 acres in extent (McKenzie and Wyndham seem to have been expanding at the same rate). The soils were light

Looking across Patrick's Plain, the vast vineyards of Alexander Munro at Bebeah. Munro's wine enterprise was managed by his kinsman William McKenzie. By W.T. Smedley, 1886, courtesy of Cameron and Jean Archer.

part-chocolate, part-sand, with limestone. The stone cellars had been extended from the first construction 60 years earlier, and were very hot. There were some large casks (this presumably points to the capacity of the business to manage a larger scale of production to make export viable). No old wine remained on site, and all of the three- to four-year-old wines were very good. Dalwood also kept a still for making brandy. The 1889 vintage appeared to be the best in many years. Nicholson listed the prices of some wines, inferring they suited his purposes: a *vin ordinaire*, and wines named as '1st red' and 'Super red' (the highest quality). Dalwood presently sold to Wood Bros in Newcastle – the city's largest brewer – but this contract was due to expire (Nicholson does not give an expiry date).

Although we have not yet mentioned John Wyndham's brother George, whose wine John sold, Nicholson visited his property, named Fern Hill, adjacent to Dalwood. Here the grape varieties were Hermitage, Verdelho, Shepherd's Riesling, Pedro Ximenes and White Shiraz. Nicholson thought the 25 acre vineyard was 'beautifully situated', the ground floor of the cellar 'good but the top floor very hot'. The vines were planted 1200 to the acre, and yields were as high as 1700 gallons per acre. Of the Fern Hill wines Nicholson tasted, he found the 1886 red good, an 1887 white 'rather tart', and an 1885 red 'light and slightly earthy'.

At Eelah, Bambach cultivated 18 acres of vines on a slope of light and loamy soil mixed with limestone. His grape varieties were Hermitage (which may have been red or white) and Riesling (Semillon), Madeira (Verdelho) and Pinot (indeterminate). Grapes were crushed with a lever press (perhaps the same one as in 1865, when a reporter visited from the *Maitland Mercury*). The cellar, constructed with double wooden walls and a bark roof, contained two 300 gallon barrels, in which the wines were presumably made from blended varieties. Nicholson stated the wines were good, especially the 1888 white, and 'worth keeping in view' to be purchased by Burgoyne; however, the 1887 was very light and watery.

It would have given Henry Carmichael and James King great pleasure to know that Nicholson called at Porphyry. The vines at Irrawang had by this time been abandoned (although for how long is not recorded), following King's death in Britain, in 1857. In 1889, there remained 20 acres of vines at Porphyry, managed by Carmichael's sons. Nicholson noted that the 'good stone cellar' with underground storage had an 'even cool temperature, and very suitable for storing old wines'. Porphyry vines lay on 'loamy, light soil nearly flat and low lying', but he left no word of wines. Kinross vineyards at Raymond Terrace were 40 acres in extent on light sandy soil on flat land. The stone cellars were good, and the wine reputed to be 'red full bodied and strong', but none was available for sale.

Other properties Nicholson visited were Andrew Doyle's Harpers Hill, with 12 acres of vines on the side of a hill with light chocolate soil, planted out mainly with Hermitage. The vines were staked and trellised, the cellar shed was hot, the vines averaged about 600 gallons per acre, Doyle charged 3 shillings per gallon for wine, and had 1000 gallons in storage. The wines had good colour and body, but tasted of seawater. At Gresford on the Upper Paterson, Edward Doyle's Clevedon Wine Cellar shared land with 25 acres of Hermitage, Verdelho and Shepherd's Riesling, and produced up to 10 000 gallons per year. Nicholson does not mention the qualities of the wine, but states there is some available for sale.

> Kinross vineyards were 40 acres in extent on light sandy soil. The stone cellars were good, and the wine 'red full bodied and strong'.

At Branxton, a settler called Dryden had a small vineyard and also bought grapes to make wine, which he chiefly sold to wine shops. There were several small growers at Harpers Hill: Holman with 10 acres of Hermitage and other varieties on a sandy slope, with a little 1888 vintage for sale; and Ryan, with 6 acres of vines, also chiefly Hermitage, on light sandy limestone soil, with 'only' 650 gallons of 1888 vintage for sale – 'coarse but sound and fair and somewhat light' – selling for 1 shilling and 6 pence a gallon.

Nicholson visited Mrs Kelman, at Orizaba, Mount View – the daughter of Edward and Susan Tyrrell, who had married into the Kelman wine family. Nicholson approved of the 1887 wine from Kelman's vineyard of 10 to 12 acres, remarking that only 500 gallons remained of that vintage, along with 4000 gallons of the new vintage. James Doyle had already bought this year's production, at 1 shilling and 9 pence a gallon.

At John Wilkinson's Coolalta, Nicholson described the 72 acres of vines as 'beautifully situated on steep hills', on soils of light chocolate and limestone. Construction was underway on a new shed to hold 18 000 gallons of wine. In 1889, the grapes were 'very small but plentiful'. Wilkinson cultivated Hermitage, Malbec, Pinot (which is not stated), Verdelho and Riesling (Semillon) – averaging at least 500 gallons per acre. Coolalta red wines were 'good but slightly earthy', and the white (presumably made into a blend), 'V. good'. Nicholson considered the 1886 red and white worth buying, at a price of 5 shillings and 6 pence a gallon. This price is relatively high, and Nicholson does not query the price compared with others, indicating that this product must have been one of the best, in the buyer's opinion, in the region.

When the Burgoyne agent called at Walter Green's property Norwood, more commonly known as Allandale for the district where it was located, he found a 55 acre vineyard (on sand, gravel and limestone), and 40 000 gallons of wine stored in large brick cellars. Green invested in collecting and maturing wines, and made a sparkling wine (which Nicholson termed 'champagne'), Riesling (Semillon) and Pinot (again, no word as to which cultivar).

Records are sketchy regarding the result of Nicholson's tour, except that from 1891 Dalwood wines were sold to Burgoyne's and shipped in barrels marked only as Australian wine (*not* as Dalwood) to London – the culture of blending and storing that the generation of imaginers had 50 years earlier sought to join. In that year Dalwood wines were also still sent to French New Caledonia, as part of a Pacific trade in Australian wines.[20]

Walter Green, Norwood (Allandale)

IN 1898, ALLANDALE VINEYARD AND CELLARS WAS A LANDMARK, 'ONE OF the most conspicuous objects that meets the gaze of a traveller on the Great Northern Railway ... fully half an acre of corrugated iron roofing with words "Green's Allandale wine cellars"' in letters visible from half a mile away, according to one newspaper reporter.[22] Green had learned his craft from his father Peter Green, and from Philobert Terrier. The racking of wine at Allandale cellars occurred with centrifugal pumping to separate wine liquid from sediment solids. Allandale wines received awards from wine competitions in Europe, India and the United States. Customers included the Austrian Imperial Army, which served Allandale wine on their warships. Green claimed that Allandale held the colony's largest stock of 'clarets and hocks', and annual production exceeded all of the wine made in the Hunter since the beginning of winegrowing.

John Murray Macdonald, Ben Ean (Pokolbin)

A LATECOMER IN THE GENERATION OF FIRST WAVE GLOBALISTS, JOHN Murray Macdonald (his choice of spelling for his surname) was born at his family's Pokolbin estate of Glenmore in 1863, and himself became a large-scale landowner in Pokolbin from the 1880s, incorporating winegrowing into his Ben Ean property. Vines were first planted at Ben Ean in 1886, and surviving business records for the farm indicate that a thriving wine

Ben Ean, Pokolbin. Macdonald Family Collection, MS A2/2–17, Box 17, Capricornia CQ Collection, Central Queensland University Library.

production existed by the 1890s. Macdonald employed perhaps half a dozen farm hands, and took on an additional 12 men for the 1900 vintage. He also purchased grapes from other local growers at vintage. Ben Ean's wine was sold by Newcastle brewer and publicans Wood Bros in Newcastle, and the Lindeman brothers and J Lynch & Co, wine merchants, in Sydney.

New South Wales' official viticulturalist Michele Blunno attended Ben Ean vintages. In 1898, he recorded sugar level tests on several wine styles in the Hunter: Hermitage, Verdot, Black Spanish, White Shiraz, Madeira

and Riesling (Semillon). These presumably included wines from Ben Ean. From these tests Blunno predicted that, as usual, Hunter styles would be light, even lighter than the previous year. The cooler season meant fewer wines tasted overheated. Blunno felt that the young white wines displayed a pleasing freshness of taste, and that some poorer quality wines could be blended with others to find a market and prevent loss of income.[23]

While Green, Macdonald, members of the Wilkinson family and Brecht weathered the onset of the 1890 Depression, others did not, and some reordering of the industry occurred in the mid-1890s. The Depression came after bullishness in foreign investment was followed by an international economic crash; this downturn crushed market prices for farm products, and many banks ceased trading or collapsed.[24] Just as new members of the Hunter wine community arrived through changes in property ownership after the 1840 Depression, the 1890s Depression provided pathways for newcomers into the industry.

John Younie Tulloch, Glen Elgin (Pokolbin)

JOHN YOUNIE TULLOCH WAS BORN AT PATRICK'S PLAINS IN 1866, WHERE his grandfather James Tulloch settled after emigrating from Scotland in 1839. John Younie Tulloch's father James became a 'leading townsman' in Branxton as store proprietor and postmaster.[25] He and his sons Alexander and John were active in local business. John also participated in politics, campaigning for Albert Gould, a Sydney-based solicitor who supported free trade and Federation, who served as Member for Patrick's Plains from 1882. John Younie Tulloch worked in his early life as a telegraph operator, and later managed a milk separator and creamery in the Branxton area. In 1891, the partnership between John and Alexander in general stores at Branxton and Greta was terminated in favour of each taking responsibility for separate stores: John at Branxton, Alexander at Greta.[26] John participated in executive roles at the Mechanics Institute – which met in the School of the Arts building at Branxton – the Branxton

John Murray Macdonald and family, Ben Ean, 1895. Back row: J.M. Macdonald (2nd left). Front row: Harriot Macdonald (left), Rev. James Benvie, Harriot's father (far right). Macdonald Family Collection, MS A2/2-17, Box 17, Capricornia CQ Collection, Central Queensland University Library.

Improvement Committee, and the local cricket club. In 1895, John Younie Tulloch joined others in petitioning for Branxton to become a municipality encompassing Dalwood and other settler properties.[27] Alexander and John Tulloch were justices of the peace, marking them as a new generation of public men, the leaders of their community.[28]

John Tulloch's acquisition of land in the heart of Pokolbin in 1895 extended the family tradition of community leadership into a new district. Tulloch received the property on McDonald's Road, Pokolbin (planted at the time with a Shiraz vineyard) in lieu of a debt to his store by a member of the Hungerford family, and renamed the land Glen Elgin. In the same period as the sons of Henry Lindeman branched out from their father's wine business base to become traders, Tulloch – the merchant – became a winegrower. This widening of roles within the community occurred in a dynamic web of connections. People knew each other, and sought to collaborate. The encounters between people in the exchange of goods and services were the basis of friendships and trust (or otherwise); they were the processes or circuitries that generated the community support that propelled winegrowing enterprises.

We can see this depth of connections in the business records of the winegrowing community. For instance, Tulloch's store at Branxton supplied Macdonald, as well as Fred Wilkinson's son Audrey Wilkinson at Oakdale. Wilkinson paid Tulloch several times for goods received (although we cannot see whether they were vineyard and winemaking supplies). Wilkinson received bread from 'JB Hungerford', beef from 'R Hungerford neighbour', and honey from Edward Tyrrell at Ashmans. He borrowed a wine press from Green at Allandale, and sold an iron plough to Harry Drayton.[29]

The movement of goods out of the region resulted from a self-generating hive of activity. Wilkinson's wines from Oakdale (which were likely produced from grapes from several properties) were sold in Sydney to at least two merchants, with a cheque for £47 received from Tooth & Co. Wilkinson also sold wine to Messrs R Metayer & Co, Wine and Spirit Merchants, 6 O'Connell Street, Sydney and Messrs Fesq & Co, Wine and Spirit Merchants, 111 Pitt Street, Sydney. Wilkinson's uncle Lionel Stephen (an insolvent after the Depression), conducted much of the transport of these wines from Oakdale by dray. Among many journeys undertaken along the roads of Pokolbin to Allandale railway station, and to other properties, Stephen took the short drive to Macdonald's Ben Ean to collect equipment including bags and wire.

Audrey Wilkinson's brother Garth travelled to Drayton's at Bellevue and the Phillips' farm for corn and sorghum seed. Supplies for Oakdale included vine stocks. Henry De Beyer provided 2500 vine cuttings from his Pokolbin 'nursery' vineyard. Within the established pattern of people's routines around the district, some wine circulated between family members and neighbours (as personal supplies, but also in some ways compensating for the lack of a formal group like the Hunter River Vineyard Association). Tyrrells bought wine from Audrey Wilkinson, and

Pokolbin, c1949. Photograph by Max Dupain. Tulloch Family Collection, courtesy of J.Y. Tulloch

Wilkinson purchased wine from family members, as well as sending samples of wine to Sydney merchants. Once wine orders were received from Sydney, large barrels were filled at Oakdale, carted to Allandale railway station by Lionel Stephen, and freighted to Sydney.

Audrey Wilkinson's diaries – and those of Macdonald – are a unique record of the bonds of close community ties and the reach of the wider circuitry of trade connections. Members of the family at Oakdale and other Wilkinson properties attended church (except when grape picking was at its height). Audrey and his siblings played tennis at Ashmans and several other farms in Pokolbin, and organised other social events. Commodity production involving people and Hunter wine is the result of a great deal of neighbourliness. The two previous generations of the Hunter wine community – the imaginers and experimenters – were bound together by habits of friendship, but relationships between many of the first wave globalists were even more tightly wound.

> Audrey Wilkinson's diaries are a unique record of the bonds of close community ties and the reach of the wider circuitry of trade connections.

In 1898, a visitor to Pokolbin described some of the patchwork of vines on mixed farms of the district. George Chick lived close to George Kiem's 16 acre vineyard, and Edward Gillard leased land from Chick, on which he tended an 8 acre vineyard. Tulloch cultivated five acres of vines. W Leaver owned 9 acres of Hermitage, Shiraz and table grapes. The Tyrrell family's vines extended across 27 acres of Hermitage, Madeira, Shiraz, Semillon and Aucerot (obtained from cuttings from Cawarra), and a further two acres of table grapes. (Mr Tyrrell, incidentally, did not believe in fortifying and his 50 000 gallon capacity cellar held 9000 gallons at this time.) Alfred Joass's farm included a two-acre vineyard. Wedged within the boundaries of Catawba, Henry De Beyer – who immigrated some years ago from Alsace with his parents, at age two – had over the past eight years greatly improved his 90 acre property, with the help of his two sons. De Beyer boasted 11 acres of vines comprised of 'Hermitage, Pineau, Shiraz, Riesling, Malbec and Madeira, besides a few choice table grapes'.

John Tulloch and his London-born wife Florence moved permanently to Glen Elgin around 1899.[30] That same year Macdonald purchased a good deal of his vineyard and winery equipment from John Younie Tulloch's Branxton store, including pruning scissors, sulphur in casks, sulphur bellows, and casks for wine storage and transport. Some of the casks purchased were coopered from cedar. Drayton family members worked for Macdonald, as did Hungerfords and Elliotts, and members of the Kiem family. Macdonald traded wine from the Elliotts, Phillipps and Hollinsheads.

Ben Ean wines were Shiraz, Hermitage, Madeira and Shepherd's Riesling, and the scale of the operation is evident in the month-long vintage at the turn of the century, extending from the end of February to the end of March. In 1900–01, expenditure on the Ben Ean cellar, 'young vines', 'old vines', stock and capital totalled just over £1281, and income from the cellar a little more than £1447 – allowing for a credit balance and a small additional debt, this represents income of close to £275. (This is approximately $35 000 in contemporary terms as income, but – rather astonishingly – a million dollars in capital.)[31] The profit appears to have been distributed between more than six family members.[32]

Macdonald left Pokolbin to become a cattle grazier in northern Queensland, reportedly because of labour shortages in the winegrowing industry.[33] He had served on Cessnock Shire Council from its inception until 1911, occupying 'a prominent position in this district for many years, and his place will be difficult to fill'.[34] Harriot Macdonald played the organ at the local Methodist Church, and she would be greatly missed. Between the Wilkinsons, Tullochs and Macdonalds as pillars of the Pokolbin Methodist community, Edward Tyrrell's Anglican connections, and the evangelical ties of the Hungerford family, the core of Hunter winegrowing in this period was decidedly Protestant, deeply enmeshed socially; a finely tuned motor of sober and industrious progress.

Macdonald, and later Tulloch, held a central role in a hub of vineyards and wineries as a major employer, grape buyer and mentor, in much the same way as earlier leaders of the Hunter wine community – James King, Henry Lindeman and John Wyndham.

John Davoren and the Dalwood legacy

POKOLBIN LANDHOLDERS WERE NOT THE ONLY DRIVERS OF THE INDUSTRY within the Lower Hunter winegrowing community. Two generations of the estate model at Dalwood, for example, had nurtured significant expertise, especially in the Davoren family.

In 1894, a former New South Wales treasurer (not named in the newspaper report) acted as host at the annual employees' dinner at Dalwood estate, attended by John Wilkinson, Dalwood manager Monsieur Gouget, senior employees (including John Davoren) and more than 80 people from the estate workforce. The speeches identified Dalwood as the second-largest wine estate in the colony (without revealing the largest), and over toasts of Dalwood wine recounted how the produce of Dalwood presently travelled to England – from where Burgoyne's transferred the barrels to France, added the wine to French products, and resold it as French wine in places such as Australia. The global disruptions affecting this generation resonated at the local core of Hunter production, and the contingencies of the globalised market created what the former New South Wales treasurer called this 'great injustice to the Australian wines'.[35]

Monsieur Gouget agreed this represented a great injustice, and that the response could only be to continue to build a reputation for Australian wine, through formation of a colonial wines trust. Gouget also claimed a rise in quality and quantity of wines from Dalwood since his arrival (at the appointment of the trustees of the estate, following John Wyndham's death). Dalwood in 1894 had 150 000 gallons of wine in storage, and the recent vintage had produced a further 25 000 to 30 000 gallons. In the continuing formalities at the dinner, John Davoren responded on behalf of the employees.

John Davoren was born at Dalwood in 1868 – perhaps within the community of workers whose cottages The Vagabond noted in 1878.[36] He and his sons would go on to be central to the New South Wales operations of Penfold's, a family wine company from South Australia. John Wilkinson purchased Dalwood in 1901, and Penfold's purchased the property in 1905 – during the next generation of Hunter winegrowing.

Taking stock

IN 1888, THE NEW SOUTH WALES STATISTICIAN COLLECTED DATA ON THE number of wine presses in the colony, and counted 360 – a third of which were based in the Hunter. Of these there were 18 wine presses in the older districts of Paterson and the Williams River, 11 at Gloucester, 52 in 'Hunter' (an official name for a part of the region taking in Maitland, Pokolbin and Rothbury), and 13 in the Upper Hunter, 6 at Morpeth, 3 in Lake Macquarie and at the southern edges of Newcastle, and 30 at Patrick's Plains.[37] Knowing that county names in this era do not align with later geographical boundaries, we will not attempt to draw conclusions that link particular wine producers to these figures. What this data *does* say is that the Hunter remained a focus of New South Wales winegrowing, and that while Victorian and South Australian producers adopted a co-operative winery infrastructure, the Hunter wine community had developed from a social base favouring individually owned presses.

For the year 1891–92, census collectors recorded the colony's vignerons, counting 2134 in total. These vinegrowers were not all winemakers. In the Hunter there were 48 vignerons around Paterson and Williams River, 3 at Gloucester (Weismantels and 2 others), 157 in the Hunter district, 40 in the Upper Hunter, 14 at Maitland, 5 at Morpeth, 17 at Northumberland, 108 at Patrick's Plains and 36 at Wollombi.[21]

Hunter wine production developed across two generations and then stabilised, as shown in the first graph in appendix 1. Across the second half of the 19th century however the concentration of vineyards moved to the west of the wider Hunter region. This is clear in the second graph in appendix 1, and the third graph indicates the dominance of wine grapes over table grapes since records began. Ultimately, since the introduction of *V. vinifera* to the Hunter in 1828, the overriding intention of those settlers who planted grapes was to make wine; and by 1900, the Hunter was unequivocally a wine region, with distinctive origins, singular characters and deepening heritage.

Chapter 7

Amid turmoil and temperance: the forgotten generation and the stoics, 1901–54

During the vintage of 1900, George Craig from Sydney visited Pokolbin and Rothbury, 'the country of the Tyrrells, Holmes, Campbells, Kings, Loves' (and of Tullochs, Draytons, Kiems, Halls and others).[1] 'The vine is seen everywhere, and so clear and healthy', laden with ripe grapes: 'Hambro, Shiraz, Hermitage, Reisling [sic], Black Prince, Malbec, Muscat, Madeira, Tokay'. Craig's guide, Hutcheson (likely AY Hutcherson), sold grapes from his vineyard near the Holmes family's Wilderness to John Murray Macdonald at Ben Ean.

Craig and Hutcheson rode on horseback along roads and across paddocks, seeing that Capper's Catawba vineyards and winery 'looked very well as I passed, and should be a good producer of the best wine grapes', wrote Craig. He saw too that 'Mr Tulloch [at Glen Elgin] is steadily cultivating large areas of vines. He also has a local creamery, which I did not visit'. Although Craig rode past the local primary school he did not visit there either. The district's children were all in the vineyards, picking grapes for pocket money. From a lookout point, Craig gazed into the heart of the country. 'Tall broken hills did not obscure range of sight.' The 'whole view' of the panorama below appeared 'arcadian … with the morning sun dancing with Ceres and Flora in the land of wine, milk and honey'.

As Craig's enthusiasm conveys, the Hunter wine community in this part of the wider region greeted the new century with a settled and confident air. On the positive side of the ledger, the first wave of globalisation remained at a crest. There were some years to run before the outbreak of World War I in 1914 curtailed global shipping, and along with it some of the Hunter's wine trade.[2] Channels of commerce remained open for wine to flow outward from the Hunter in barrels (and sometimes bottles), not only by rail to Newcastle, Sydney and within New South Wales, Australia and the Pacific – but also in a small degree to Britain. Flowing inward from Europe and elsewhere were new technologies encouraged by the state viticultural expert, some specialist equipment such as tools, concrete fermenting tanks, and the inputs required from beyond Australia: oak barrels (often from the United States), bottles, and sulphur to treat pests and diseases as well as to use in winemaking.

Edward Capper was proud of the modern equipment in his winery in 1903. E.P. Capper and H.H. Capper, Catawba Wines, 1903, State Library of NSW, ML 663.2/7A1.

TOP: Catawba grapes are a hybrid of American *Vitis* and *V. vinifera*. The property was named for an American poem about Catawba. E.P. Capper and H.H. Capper, Catawba Wines, 1903, State Library of NSW, ML 663.2/7A1.
BOTTOM: The barrel room with concrete vats for fermentation and racking. E.P. Capper and H.H. Capper, Catawba Wines, 1903, State Library of NSW, ML 663.2/7A1.

Knowledge circulated too, in the continuing development of the international grape and wine sciences: new methods of vine cultivation and winemaking, new forms of expertise – often gained from Roseworthy College in South Australia, and more rarely from Europe. Wine competitions proceeded not only at local, state and interstate agricultural shows, but internationally – and Hunter winegrowers achieved outstanding results beyond the region. New South Wales was the third largest wine-producing state, well behind production levels in South Australia and Victoria, but ahead of Queensland and Western Australia, with New South Wales production focused in the Hunter, the outskirts of Sydney and the Riverina.

Between 1901 and the early 1920s there were no devastating weather events or new pests or diseases to slow the impetus of wine production in the Hunter. A gathering disapproval of the social harm caused by drunkenness did not yet affect the wine industry. Even so, 1901 brought a monumental shake-up unique to Australia in the global wine industry: Federation of the British colonies of Australia into a newly constituted nation of states with the same borders as the former colonies (and maintaining separate state legislatures nested within a new upper layer of federal government). Federation is generally viewed as an anticlimactic form of nationhood – lacking drama to inspire pride, compared with, say, the bloody birth of the United States from the American War of Independence. Students of Australian history tend to stifle yawns as they learn that the impulse for federated nationhood focused on the geographical determinism of one island continent, one nation – and on administrative motivations such as the removal of intercolonial tariff barriers, and standardisation of a national postal service.

Yet, for the South Australian wine industry, if not for the other new wine states, Federation spelled new markets. South Australia's Thomas Hardy foresaw, with elation, that when his colony became a state within the new Australian Commonwealth, the wide range of regional vineyard locations, such as McLaren Vale and Barossa, would offer an appealing diversity of wine styles – from dry Burgundies through to heavier fortified styles – to drinkers in other states.[3]

By this time, while grape growers spoke to one another in terms of *grape varietals* (although Madeira continued to be known by place of origin, rather than as Verdelho), winemakers spoke to each other and to consumers in terms of wine *styles*, determined by strength of alcohol and sweetness (or body), in two classes. 'Dry' wines were Burgundy (a heavy white), Hock (medium-bodied white, made from grapes such as Semillon or Verdelho), and Chablis (a lighter white). Reds were clarets (which might be any red grape), or heavier Burgundies (any red grape with a strength of alcohol between claret and port). A Burgundy could be either white or red, but in each case meant a heavy-bodied wine containing alcohol. 12.5 per cent alcohol by volume (ABV), or 25-proof. The 'sweet' class of wines comprised port, Muscat, Madeira and Frontignac – each of over 15 per cent ABV. Sherries were the only dry wines that were fortified with one-third of (grape) brandy. 'Sweet sherries' were similar to a port style. 'Sauternes' were made from late-harvested white grapes with some unfermented grape juice added to boost sweetness. Australian 'dry' wines were usually 10 to 12.5 per cent ABV, contrast with European wines, which contained 8 to 11 per cent ABV.

The 'sweet' class of wines comprised port, Muscat, Madeira and Frontignac – each with 30 per cent additional alcohol. Sherries were the only dry wines that were fortified with one-third of (grape) brandy. 'Sweet sherries' were similar to a port style. 'Sauternes' were made from late-harvested white grapes with some unfermented grape juice added to boost sweetness. After fortification for export Australian 'dry' wines could be at 20 to 25 per cent pure, in contrast with European wines, at 16 to 22 per cent alcohol. The Hunter made chiefly Hocks and clarets. Other New South Wales regions made other styles (heavier wines in the warmer Riverina and sparkling wines near Sydney). South Australian wines were chiefly Burgundies, or heavier styles. Prior to 1901, the existence of some colonial border taxes, and the low degree of interest among colonists in wine from beyond their own borders, meant little intercolonial wine trade.[4]

Compared with their counterparts in South Australia, Hunter winegrowers were anxious about an inundation of products into their traditional markets, an anxiety that barely registers in the broader understanding of

The romance of turn of the century vintage is captured in this promotional photograph for Catawba Wines in 1903. E.P. Capper and H.H. Capper, Catawba Wines, 1903, State Library of NSW, ML 663.2/7A1.

Federation. This is one of the reasons we have referred to the fourth generation of the Hunter wine community as 'forgotten'. Although several of the names of winegrowers from this era are known today – such as Audrey Wilkinson, Dan Tyrrell, John Younie Tulloch's son Hector and Maurice O'Shea – this fame is the result of the continuity of family businesses and interest in prominent past characters. From 1901 to 1929, the work of Hunter winegrowers, and the fate of those who left the industry during coming troubles with vine disease, grape oversupply and temperance laws, was buried and forgotten under the larger idea of the Australian nation and the weight of wider world events.

Within the broader Hunter region, the national steel industry emerged in Newcastle in 1915. Soon, the Hunter hinterland – once so full of promise compared with the coastal entrepot to the Hunter River – was surpassed in economic and social importance as the port township grew into a city from the late 19th century. In the first half of the 20th century, coal development crept westward, outwards from the mining villages once surrounding Newcastle. As hinterland coal seams were discovered, new townships of miners formed – centred on the town of Cessnock near Pokolbin and Rothbury – that were chiefly working class and beer-drinking, with almost no wine-drinking middle class to speak of.

We have called the following generation, from 1931 to 1954 (the

Drayton, Elliott, Steven, Trevena, Tulloch, Wilkinson families, and employees of the McWilliam's, Penfold's and Lindeman's wine companies), 'the stoics', as stoicism is a philosophy that encourages duty over profit. These grape growers and winemakers made heavier styles to maintain an income, and continued to refine table-wine styles for a small but committed consumer base – and they waited through the understandably sombre mood of a nation bearing the effects of World War I, the Depression, World War II, and the threat of a third world war during conflict in Korea in the early 1950s.

Loading barrels. E.P. Capper and H.H. Capper, *Catawba Wines*, 1903, State Library of NSW, ML 663.2/7A1.

Hocks and clarets, deluged by new wine

ALTHOUGH THEY WERE ALL PART OF THE BRITISH EMPIRE – AND contained within a single island continent – until 1901, Australia's wine colonies were really separate wine countries. For a start, there were nuances in the nature of their settlement, and in approaches to politics, production and trade – including diverse attitudes to trade protection. New South Wales and Victoria were founded with an unfree convict labour workforce prone to seeking consolation in drinking alcohol. These states had histories of close attention to controlling alcohol production and trade to prevent drunkenness. In South Australia, which did not have a convict labour force, and relied instead on attracting free settlers (who were often church-minded), a more restrained drinking culture emerged. Victoria instituted protective tariffs to support its wine industry, whereas New South Wales and South Australia were free traders. Subtle differences also existed in the composition of the ethnicities and occupations of these colonial societies. Although British settlers were the principal winegrowers, in each of the wine colonies there were differences in the distribution of immigrants from Europe during the 19th century. As well as the Germans from the south-western Rheingau in New South Wales there were immigrants from eastern German states in South Australia. In the central western district of New South Wales, a Croatian family began winegrowing in Young. Victoria's winegrowers included a Swiss–French population. And, across the whole, was a sprinkling of Italians and a few French.

Within the Australian wine colonies that became wine states there were diverse wine regions with different geographies of soils, diurnal temperatures and prevailing winds – shaping the vinegrowing ways of the different communities who were cultivating grapes, and making, selling and drinking wine. The emergence of separately governed British Australian colonies during the 19th century also meant that in 1901 new Australian 'wine states' were natural economic competitors, not collaborators. As historian David Dunstan puts it, because South Australia possessed the largest colonial wine industry by volume – the 'new "national" wine industry' increasingly became 'a South Australian preserve with ownership concentrated in South Australia'.[5]

While the tide of South Australian wine flowing to New South Wales rose, the New South Wales press predicted great success for 'the hocks and the clarets of the Hunter River' in the new national wine market – 'for no part of this huge continent of ours can grow this particular description of wine to greater perfection than the Hunter River vignerons'.[6] New South Wales viticultural wine science expert Michele Blunno pronounced Hunter wines to be 'of a superior class', pointing to sales secured in Victoria and South Australia and plans to newly test the English market. Within the wider reach of the new national industry climate, Blunno approved of growing interest in the Hunter for the creation of a regional association that would represent the small region in larger organisations – and which would channel communication to the three levels of government that now needed to be negotiated: local shire councils (or municipal level), and state and national governments.[7]

Each level of government funded or regulated different factors with direct bearing on Hunter winegrowers, from council (or municipal) and state management of roads, to state jurisdiction over freight and transport, licensing of premises to sell alcohol, and extension services such as Blunno's employment. The new federal government also oversaw policy on exports.

The Pokolbin and District Vinegrowers Association

FOR HUNTER WINEGROWERS WHO SAW THE NEED FOR A COMPASS TO navigate the confounding terrain of new administrative structures, a collective regional approach seemed advisable, as a means of ensuring that processes affecting the Hunter wine community developed in a manner in which they had a direct say.

In the late winter of 1901, pruning progressed in the Hunter vineyards, vine plantings were extended with new and recommended varieties, cellars were enlarged, and a feeling of confidence prevailed.[8] Still, the *Maitland Daily Mercury* (successor of the earlier *Maitland Mercury*) called pointedly

Pokolbin and District Vine Growers Association, year unknown. Tulloch Family Collection.

for a regional branch of the Australian Winegrowers Association.[9] This public urgency reflected conversations within the community. In September, winegrowers gathered at Pokolbin Hall to formalise the new Pokolbin and District Vinegrowers Association (PDVGA) – the first such organisation for the region since the demise of the Hunter River Vineyard Association (HRVA) in 1876.[10]

Inaugural members of the PDVGA were a rollcall of the public men of Hunter wine – many of whom we have introduced, some of whom are less visible in the broader historical record of the Hunter wine community.

The PDVGA began initially as a larger group than the earliest incarnations of the HRVA. Those present at the first meeting of the PDVGA (and elected to positions within the organisation) were president Stewart Corner (of Ivanhoe, Pokolbin), and vice-presidents Charles Holmes, John Younie Tulloch, George Brown (Marthaville in Cessnock) and A Harman. Audrey Wilkinson took the role of honorary secretary and John Murray Macdonald that of honorary treasurer. Others in attendance were Edward GY (Dan) Tyrrell, William Wilkinson, Lionel Stephen, J Love, S Holmes, Andrew Kelman, WJ Macdonald, Charles Wilkinson, AY Hutcherson, GA Campbell and George Hall.[11]

The PDVGA aimed to promote Hunter wine exports internationally, to ward against loss of custom through competition with South Australia, and to provide a collective voice on behalf of the region as the scaling up of the industry to a national level made concerted regional identity more important in suddenly vaster territory. The association's concerns were similar to those of the HRVA, but instead of conjuring an industry – as did the imaginers and experimenters – the PDVGA faced the decline of their bottom line.

Early matters related to wine quality. The *Wine Prevention Act* (1902) regulated the quality of wine in New South Wales by prohibiting any additives (sugar, essential oils, cherry, barium, arsenic, salicylic acid – and later saccharose – and so on), and capped the addition of grape spirit to 25 per cent for dry wines, and 35 per cent for sweet wines. But policing of the problems of adulteration proved feeble, and the PDVGA and Audrey Wilkinson, especially, are credited with pushing Minister for Mines and Agriculture John Kidd (reportedly an apathetic public servant) to enforce the legislation, through inspection of wares and the naming of brands that were doctored.[12] The act gave protection too against wrongful branding of wines – a critical measure to safeguard the reputation of Hunter wines against fraudsters capitalising on the flux in the market resulting from the inflow of wines from other states.

The PDVGA did not have the wide public profile of the HRVA of the previous century. Its membership comprised key members of the Pokolbin agricultural community who continued to attempt to influence government

PREVIOUS PAGES:
Empty wine barrels, likely from Lindeman's Cawarra. The Wharf, Gresford Crossing, Paterson River / photographed by Kerry & Co. ca. 1895, State Library of NSW, Call no. SPF/2690 SPF/2691, SPF/2692.

policy. Yet the PDVGA became subsumed into the Pokolbin Agricultural Bureau led by John Younie Tulloch – an organisation for all of the district's producers, most of whom were involved in more than one form of production to weather the uncertainties of farm income from year to year.

Companies rising

BEFORE FEDERATION, THE NEW SOUTH WALES PRESS COMMONLY favoured the achievements of New South Wales wines. After 1901, along with the torrent of wine flowing into Sydney across half a continent from the Lower Murray River Basin, came South Australian wine companies – to sell wines in Sydney, and to invest in New South Wales wine regions. The traditional bounds of New South Wales wine business shifted as a result, and the Sydney press soon adjusted allegiances. When, in 1905, Penfold's purchased Dalwood, the *Sydney Morning Herald* laid out a welcome mat in breathless prose. The removal of intercolonial tariff boundaries, reported the newspaper, had the great benefit of bringing Penfold's table wines and brandies of astonishing quality and value – and of fame in Britain, India, Japan and China – 'prominently before the New South Wales public'.[13] Penfold's won many gold medals in competition in Paris, and a jury appointed by the king in London 'pronounced them as the best from Australia'.

Meanwhile, in 1907, Lindeman's Ltd, the company born of Henry Lindeman's enterprise at Gresford – and now managed by his sons Arthur (based at Cawarra) and Charles (in Sydney) – created an ostentatious display. Sixty hogshead-sized barrels of Cawarra wines were transported on draughthorse-drawn 'lorries' in procession from Lindeman's Cawarra cellars in the Queen Victoria Building in Sydney's George Street, to Circular Quay, to be sent by mail steamer to London.[14] Lindeman's operated one of the largest wine cellars in the world, and the UK market offered a destination for this company's wines during a new competitive climate resulting from South Australian products.

The dual domination of Penfold's and Lindeman's at the outset of a new national wine industry is evident in the results of a 'colonial wine' competition in London in 1908: across eight classes, of 19 awards received (all by Australian entrants), Penfold's won a gold and a bronze medal, and Lindeman's two silver medals.¹⁵ Another New South Wales producer, from outside the Hunter, received four medals. At the Royal Agricultural Society's Great National Exhibition in Sydney in 1909, the outstanding wine competitors were Lindeman's, Penfold's and Thomas Hardy & Sons.¹⁶

The dynamism of the period is plain in corporate jostling. Penfold's had bought into the Hunter in 1905, acquiring Dalwood. In 1906 Lindeman's incorporated, and in 1910 purchased 160 acres of 'phylloxera free' land in the Hunter, where the company intended to extend grape plantings for Hocks, clarets and 'Chablis' (a French region producing fresh, dry white wines), while importing phylloxera-resistant rootstocks.[17]

In 1911, of the principal wine firms of the state, more than half were tied to the Hunter, including G & J Carmichael, A Kelman, E Capper & Co and Audrey Wilkinson & Co.[18] (Within a few years, the properties of all of these firms, except Wilkinson, would belong to Lindeman's.) In 1912, Lindeman's Ltd purchased the Ben Ean vineyards, winery and distillery at Pokolbin from John Murray Macdonald. Macdonald's relocation to a cattle station in north Queensland provided the Lindeman company with an entrée to Pokolbin – the first of the larger companies to do so. The purchase gave Lindeman's the mantle of largest winegrowers in the district, with 90 acres of vines, and there were 20 other growers with vineyards ranging from 9 to 45 acres in extent.[19]

In competition in London in 1913, Lindeman's wines – from nine entries – won six gold medals and three silvers, for styles branded as Champagne, claret, Dry Burgundy, port, Hock and Chablis (all from the Hunter, apart from the port, which was made from Lindeman's vineyards at Corowa, near Albury, in the south of the state).[20] The company's 1000 acres of vines extended from the original Cawarra site at Gresford (still containing a small acreage), to Pokolbin and Corowa.[21] In 1914, Lindeman's purchased Kirkton from the Kelman family, with 50 acres under vines – and in press reports subtle but misleading links began to be made between Kirkton and James Busby that concealed the heritage of the Kelman family.[22] By late in 1914, the Lindeman's wine company had folded the Kirkton legacy into its own, now purporting that the company's earliest vine plantings had been at Kirkton in 1830.[23] Later, the Lindeman's story would be untangled from that of Kirkton, but the pattern woven at this time remained in records that contributed to mistaken histories of Hunter wine.

Within a few years, Lindeman's purchased Coolalta (formerly a Wilkinson family property) and Porphyry from the Carmichael family

Barrels of Lindeman's Cawarra wines on carts bound for London. Charles Kerry photograph from Allan A. Hedges collection. Sydney Reference Collection SRC18026, City of Sydney Archives.

– drawing together names associated with the imaginers and experimenters of Hunter wine under a corporate canopy. Unintentionally, perhaps – but significantly nonetheless – the Lindeman company's purchase of historic properties prolonged the presence of the names Kirkton, Coolalta – and particularly Porphyry (of later Porphyry Pearl fame) – beyond the life of the vineyards and wine families originally attached to them.

In 1933, Penfold's leased the Hunter Valley Distillery and its vineyards, managed by Charles Holmes until his death in 1928 – and in 1948, Penfold's purchased the distillery site.

Penfold's and Lindeman's were by now among Australia's leading wine companies.

One of the most poignant symbols of the connection between earlier generations of winegrowers, and the behemoths that the large wine companies would become, is hidden from the view of contemporary visitors to the Hunter wine region. Within the hilly folds of Gresford, St Anne's church and churchyard are enduring remnants of the Lindeman family story: several sturdy, sombre, grey headstones dispersed within the quaint cemetery memorialising Henry Lindeman and his descendants. The cemetery surrounds are redolent with the air of an English landscape, and could equally be tucked into a corner of the Lakes District. Inside the tiny church there is no mistaking the joyous Australian light streaming through leadlight windows onto golden- and red-timbered pew-ends carved with grapes (and tobacco leaves). A church window in memory of Arthur Henry Lindeman – who died very suddenly in hospital in 1911 – features an image of Christ, and the Gospel verse, John 15:1, 'I am the true vine'. A small image of Cawarra homestead adorns the upper left side of the frame. Although the iconography of vines and wine echoes Scottish Enlightenment-era veneration of grape (and olive) imagery, the carvings and window appear as expressions of community endeavour and of family devotion to a suddenly lost Lindeman loved one – as well as to the family's devotion to winegrowing.[24] A record of the

> Within a few years, Lindeman's purchased Coolalta and Porphyry – drawing together names associated with the imaginers and experimenters of Hunter wine under a corporate canopy.

grave-sites of past members of the Hunter winegrowing community is included in appendix 2. The Wilderness Cemetery at Rothbury is another of the community's most significant sites of remembrance.

Guiles of government, guiles of nature

IN 1906, RHAPSODISING IN A NEW SOUTH WALES GOVERNMENT publication for prospective immigrants, New South Wales viticultural expert Michele Blunno commented that the state's winegrowing – exemplified by the Hunter and Albury – had reached maturity. Table grapes were by now a third of the state's vine crop, but wine produced from the remaining two-thirds made the Hunter, especially, an antipodean idyll. 'Vignerons in New South Wales', wrote Blunno, 'although not forming a very large community, make a nice living, and this is borne out by the look of their homesteads, the happiness in their family relations, and the easy life that they enjoy'. He continued:

> I have had the privilege of visiting most of the vine-growers of this State, and always left edified with the appearance of everything, and with their amiability and hospitality, and even, when I examined their position from a critic's point of view, I could see that it was not show, but substance.[25]

'The progress made during the last few years is enormous', Blunno enthused; 'the seeds of knowledge fell on good ground; the keen capacity of the vine-growers for seizing information and turning it to practical account has been to me always a source of delight'.[26]

In 1897, Blunno had recommended that the New South Wales government discontinue controversial policies of on-vineyard treatment of phylloxera in infected districts, and that it should cease costly compensation of grape growers who were required to pull out their vines when treatments of phylloxera failed. Instead, Blunno instituted a system of viticultural stations to grow phylloxera-resistant *Vitis* species for grafting with *V. vinifera*. These plant nurseries were built at sites including

Howlong in the Riverina, briefly Raymond Terrace in the Hunter (1909–12) and later Narara on the central coast.

In 1912, Blunno departed the New South Wales Department of Agriculture amid criticism from Under Secretary of Agriculture in New South Wales George Valder, and others, of the ponderous and costly nature of the viticultural nursery program for phylloxera-resistant rootstocks, and the benefits to 'wealthy winegrowers' of subsidised vine nurseries.[27] With the replacement of Blunno as viticultural expert, first by HE Laffer (1912 to 1920) and then HL Manuel (1920 to 1955) – both from Roseworthy – Department of Agriculture focus increasingly moved away from the Hunter. Already, prior to Blunno's replacement, Valder saw the need for other areas of the state to produce 'good quality wines' in competition with the Hunter.[28] From a statewide perspective, this seemed beneficial. For the Hunter, competition threatened to dilute a hard-won reputation.

Manuel's tenure as viticultural expert began as grape growers detected downy mildew (a disease now classified as an alga), which attacked *V. vinifera* after arriving in Australia on stocks of phylloxera-resistant vines imported from France. The Hunter seemed to be especially plagued by this disease.[29]

Powdery mildew, being white, is readily visible on leaves, but downy mildew appears as angular yellow spots emerging from within the tissue of the vine leaf and is not as readily detected, which may have led to delays in treatment by some growers. Treatment also added to the cost of vineyards. Audrey Wilkinson recommended treating vine mildews with a mixture concocted from 70 pounds of 'black sulphur', 30 pounds of lime and a pound of Paris green (copper acetate), which also acted as an insecticide. Wilkinson would spray his vines up to eight times in the growing season, whether any disease showed or not.[30]

In 1924, devastation from downy mildew greatly reduced the state's wine production, but the 1925 vintage proved to be unthinkably poor in the Hunter – with three-quarters of the crop lost to downy mildew combined with hail storms. Given that the wider climate of the industry already seemed inexorably bleak, this event proved a tipping point for many Hunter vinegrowers, who chose to leave the industry, pulling out their vines and turning their land to other uses. But as one New South Wales wine producer remarked at an industry meeting in 1922, 'there was a greater fungus than downy mildew – and that was Prohibition'.[31] This producer feared the Australian alcohol industry might face the same ban on alcohol production, importation and distribution that occurred in the United States from 1920 to 1933.

Temperance laws, policy flaws

DURING THE ECONOMIC, SOCIAL AND CULTURAL INSTABILITY OF World War I, a surge in temperance values emerged in Australia and the wider Anglo world from among a tangle of different agendas. Feminist advocates of temperance, for instance, rightly sought an end to the suffering of women and families living with the drunkenness of husbands and fathers. Some temperance advocates sought complete abstention from alcohol as an expression of moral restraint – and therefore moral authority – and as a measure to improve the health of national bodies and minds.

In 1916, the New South Wales government (with jurisdiction for licensing the sale of alcohol) held a non-compulsory referendum, in which half a million people voted on the closing time for premises that sold alcohol. The options for closing time were each hour from 6 pm to 11 pm. The 6 pm option had wide support, not only from church groups and women's temperance organisations, but also from unions and the press.[32] Overwhelmingly, two-thirds of voters selected six o'clock closing; nine o'clock placed a very distant second.

The new regime of six o'clock closing began on 21 July 1916, and was legislated to remain in place until at least six months after the end of the war (with no notion yet of when that might be). The six o'clock rule applied to the drinking of all intoxicating liquors in any place licensed to serve alcohol – pubs, wine shops, clubs, hotel restaurants, railway refreshments rooms and rail dining cars.[33] While far more people were affected by the earlier closing of pubs, places for drinking wine outside of private homes were also affected.

Between 1907 and 1917, 344 hotels across the state were closed. These ranged from working men's pubs to high class establishments. Also closed were 58 wine shops, and clubs began to shut down as well.[34] (New South Wales legislators did not make provision for wine drinking at public venues after 6 pm, as occurred in the other wine states.) The Hunter's reputation for outstanding table wines remained undiminished,[35] but there were fewer and fewer places where people could drink these wines as part of a fine dining culture or wine connoisseurship. Humble wine-drinking habits were catered for in a smaller number of wine shops, less convivially and more furtively.

Opportunities to sell Hunter wine also diminished at the global scale when avenues of maritime trade and communication were rapidly closed from 1914, following Britain's declaration of war against the Axis powers led by Germany. As Australia's foreign policy remained tied to Britain's (until 1942) despite federal nationhood, Australia was also at war. This conflict affected more than international trade, it recast the status of German-born Australians and their offspring in Australia – along with visiting citizens of enemy nations. Australian authorities interned close to 7000 Germans and German descendants as the 'enemy at home', causing tremendous suffering, especially among those for whom Australia was their only home.[36] Many Germans and German-Australians concealed their non-British heritage during the war, and not until after World War II did the veil of anxiety about Germanness lift, as Australia again welcomed them in new waves of mass migration.

> In 1923, the phrase 'Australian wine' replaced 'colonial wine' in New South Wales licensing documents.

In 1919, after the Armistice that ended World War I, the New South Wales government revisited temperance laws, leading to a further reduction in the number of places serving alcohol.[37] In ensuing years, despite a growing population in New South Wales, an increasing number of licences to serve alcohol were voluntarily cancelled.

In 1923, the phrase 'Australian wine' replaced 'colonial wine' in New South Wales licensing documents, to differentiate the wines sold in wine shops from imported styles.³⁸ (Wine and spirit merchants remained the purveyors of imported wines.) The Hunter wine community continued to hope that the light, dry table wines of the region might become the national drink, rather than the excessively sweet wines that dominated production in some other regions.³⁹ In 1925, John Younie Tulloch achieved best wine exhibitor at the Maitland Show – the year of a subtle but important shift in the practices of show judging. Manuel the viticultural expert signalled that he could no longer attend this regional event, as he could not be spared from his other duties (as a sweet wine expert, perhaps Manuel found his expertise wanting in the Hunter – and he definitely considered the Hunter a region destined for an inevitable demise). Manuel suggested, however, that competitors at the Maitland Show taste all of the entries, to understand other wines and improve their products.⁴⁰ This spurred a lasting tradition. It is also notable that local, regional and state level shows (such as the Royal Agricultural Society of New South Wales Easter show in Sydney) became the principal benchmarks for many producers – rather than imported wines. This established an internal dialogue in taste and style, for a smaller and more localised market, that would result in future distinctiveness.

By 1928, very few registered clubs were in operation. A publican's licence cost £500 per year, and renewals were 5 per cent of the gross cost of liquor purchased by the licencee. Pubs could open between 6 am and 6 pm, except on Sundays, Good Friday and Christmas Day. An Australian wine licence entitled the holder to sell wines of an Australian colony of less than 35 per cent proof spirit, in quantities of less than 2 gallons – from 7 am to 6 pm – with the same restrictions on Sundays and religious holidays as for pubs. Wine licences cost £50 per annum, and renewals were charged at 2 per cent of the gross value of liquor purchased for the premises. Wine could be purchased to be consumed on or off the premises. Wine and spirit merchants could sell liquor in quantities greater than 2 gallons but it was not to be consumed on the premises. These licences cost £30 in Sydney and £20 elsewhere, and renewal charges were equivalent to those for wine shop licencees. Vignerons could sell their own wine of any quantity directly to customers, from the winery – to be carried away for consumption.⁴¹

What this summary of liquor laws does not capture however are the social and cultural nuances of drinking, such as the prohibition of musical performances and dancing in all licensed premises, which had been in force since 1898. Significantly, too, from at least 1898 licencees had been forbidden to serve all Aboriginal adults. This is an important point to underline. Although many Aboriginal people were not only visible but also crucial members of the wider early colonial Hunter community that included winegrowers, by the end of the century they were erased from the drinking as well as the growing and making phases of Hunter winegrowing. As anthropologist Maggie Brady has recently shown, there is a hidden history of government policies of 'proper drinking' for Aboriginal people, which in fact continually compounded their exclusion from settler drinking cultures.[42]

Soldier settlers

DURING THE WAR, WHEN RESTRICTIONS ON MERCHANT SHIPPING LIMITED the importation of brandy and sweet wines, Australian grape growers were unable to meet demand from winemakers and distillers, leading to encouragement to grow grapes to make brandy. Also after World War I, irrigation infrastructure pioneered in the Riverina region of southern New South Wales was extended to parts of Victoria and South Australia. Consequently, when the federal government implemented new polices on Closer Settlement from 1919 – providing returned servicemen with generous terms for small farm acreages in gratitude for fighting abroad – many hundreds of these soldier settlers were urged to become grape growers, and those in irrigated wine regions were encouraged to grow Doradillo grapes to produce brandy.

At least 80 soldier settler blocks were offered on former Crown land suitable for 'viticulture & hogs' in the western portions of the Hunter Valley, and some others on former private land. Sixty-five of these sites available to settlers were in the Fordwich Soldier Settlement, immediately west-northwest of Pokolbin.[43] These blocks did not receive any water apart from rainfall.

In Pokolbin, Hector Trevena's father received a soldier settler block on Hermitage Road in 1919 or 1920, planted Semillon vines and operated a small dairy. The grapes were sold to Elliott's at Oakvale (as distinct from Wilkinson's Oakdale). A mere 2 miles distant, Oakvale seemed 'a long way' when this trip was taken half a dozen times a day during vintage, with a dray of three hogsheads laden with grapes. Hec remembers that pickers were paid a penny per kerosene tin to harvest the grapes.[44] (Hec and his daughter Trudy-Ann continued to produce highly sought-after grapes at Trevena vineyards until his death in 2014, aged 85 years.)[45]

By 1927, several of the Fordwich soldier settlers with dairy farms and vineyards were experiencing hardship. Those with sandy soil on their blocks were struggling to make a living. One grape grower received only £204 for his grape crop, from 10 acres of vines, which was insufficient income for the year.[46]

The travails of the Fordwich settlers were overlooked at the national scale due to the disaster playing out for Doradillo growers in irrigated districts. As soon as the Doradillo vines came into bearing in 1924, amid declining demand for brandy and wine, grape prices fell. To compensate, the federal and state governments subsidised growers of Doradillo grapes, and paid a bounty on the export of fortified wines containing Doradillo brandy.[47] While returned servicemen in the Hunter struggled with the arrival of downy mildew – as did the old timers – soldier settler grape growers outside the Hunter region were also assailed by phylloxera. In 1925, settlers with infected vines were granted reduced rates of repayment on their land.[48]

Soldier settlement was not considered a success. Most settlers left their blocks because of their lack of experience and the poor quality or location of the land they were granted. Settlers who remained on the land were often those who had expanded their holdings, as did the Trevena family (who count among the few soldier settler success stories).[49] Historian Bill Gammage found that those soldier settlers who remained long enough to gain an understanding of local conditions survived by finding 'local remedies for local problems', as had Aboriginal people before them.[50]

Members of the Tulloch family purchased several soldier settler blocks when the original grantees left farming.

Maurice O'Shea, Mount Pleasant (Pokolbin)

MAURICE O'SHEA ARRIVED IN POKOLBIN LATE IN THE 1920S, A RARE newcomer – neither soldier settler, nor part of a large wine company. Maurice's Irish father John held a wine licence in Sydney's CBD at the time that Maurice was born in 1897. After John O'Shea's death in 1912, Maurice's French mother Léontine assumed control of the family business, the New South Wales Wine and Spirit Company. In 1914, Léontine sent Maurice to stay with relatives in Montpellier, France, and he completed his secondary education at Montpellier High School in 1916. In 1917, Maurice proceeded to the École d'Agriculture de Grignon in northern central France to undertake a two-year diploma in 'agricultural engineering'. He also completed an internship in the agricultural chemistry laboratory École Nationale d'Agriculture, in Montpellier, and then returned to Australia after the war.[51]

Maurice O' Shea. Photograph by Max Dupain, [Testing the fermenting grapes, Mount Pleasant winery, New South Wales, 1950] [picture]. National Library of Australia, PIC/8729/20 LOC Album 757.

In 1922, Léontine O'Shea purchased 80 acres of land at the foot of the Brokenback Range a few miles from Ben Ean and Glen Elgin, from the King family, including some acres known within Pokolbin as King's vineyard. This timing thrust the O'Shea family into the midst of the oncoming crisis in the Hunter and the wider Australian wine industry, yet – through innovations in grape blending, from his own vineyard and many others in the district – O'Shea brought a new vigour to the district. He joined the PDVGA, and readily engaged in the politics and society of his peers in the community.

At the 1924 Cessnock Show, O'Shea excelled in the port and sherry classes,

Cutting the cake after pressing grapes. Pokolbin, c1949. Photograph by Max Dupain. Tulloch Family Collection, courtesy of J.Y. Tulloch.

along with John Younie Tulloch, while Kelman and Martindale entries won for dry wines.[52]

In 1932, O'Shea incorporated his family's assets as Mount Pleasant Wines Ltd. The McWilliam wine company, based at Hanwood in Griffith, purchased a half-share of the new company, and appointed O'Shea as manager; this investment allowed new vine plantings. In 1941, McWilliam's bought the final share of the business from O'Shea, and by 1946 had acquired Lovedale and Rosehill vineyards, to create a widening patchwork of deepening ties to certain sites for vines.

Hunter locale

WHILE THERE IS NO STUDY OFFERING A HISTORICAL GEOGRAPHY OF GRAPE growing in Australia, DN Jeans has stated that by 1901 the forms of agricultural production in New South Wales had undergone sufficient experimentation for the state to have recognisable cultures of cultivation – or habits arising broadly from cropping and grazing – suited to the natural environment. Jeans, who does not discuss grape growing, does indicate that those farming systems 'made up of holdings, technology, transport routes and vehicles and ancillary services' had resolved puzzles about the qualities of 'soils, climate, water supply, terrain … and proximity to markets' and that local traditions had emerged. New South Wales farming systems were distinguished, for instance, by a history of labour scarcity, which meant that 'labour-saving methods were prized'.[53] At Pokolbin, the tradition of emptying out the school to allow students to assist with the vintage (as Craig observed in 1900 and which continued at least until the 1920s) may seem strange to contemporary minds, for example, but such rituals shaped and reshaped generations within the community. Other traditions grew from the types of grape cultivars shared and planted, and the best methods of gaining quality grape yields.

Audrey Wilkinson had managed his family's property with vineyard, dairy, pigs and other products from the age of 15, in 1892.[54] His presentation on winegrowing at a regional agricultural conference in the Hunter in 1926 demonstrates the acute rapport with Hunter vine and wine production traditions that Wilkinson had gained from practising his craft. 'Flat sandy soil produced white wines best and hilly red marl [limestone clay], chocolate and porous country the best red wines', he told his audience.[55] 'Soils should be prepared by ploughing at least three times; one of these times should be to a depth of 18 inches.' This great depth of ploughing resembled the trenching of earlier times.

Wilkinson continued, stating that the 'vineyard should be laid out with the vines planted six feet apart in rows, and the rows eight feet apart'. To mark out the spacing between vines and between rows, Wilkinson advised the use of a frame made from wire lengths soldered with

Pokolbin, c1949. Photograph by Max Dupain. Tulloch Family Collection, courtesy of J.Y. Tulloch.

cross-pieces at the required widths. The clean parallel lines of vineyards were achieved through careful planning for the continuing needs of vine management – space for the horse-drawn plough to move between the vines, room for vine canes to extend in each direction along espalier wires between posts, in order to produce healthy grape bunches, and a suitable yield. (The ideal yield from a vineyard is few enough grapes to ensure good quality fruit, and enough of that fruit to make a living.)

To replace exhausted vines after some 25 to 40 years, new plants could be propagated in the Hunter from existing vineyards without the need for phylloxera-resistant rootstocks – enabling further plant acclimatisation. To grow suitable vine clones, 'Cuttings should be carefully selected from bearing wood only, always choosing clean and well grown vines, to supply them, and if possible, they should be taken from poorer soil than that in which it is intended to plant them'; cuttings were preferable because vine plants 'were always changed by transplanting'. Each 'cutting should be about 14 inches long, and have well grouped buds. They should be planted upright, and the soil pressed well down'. Wilkinson preferred to plant new vines in June: 'one good watering may mean the difference between a good strike [of plant growth] and a complete failure'. Once vines struck or sprouted, all but the top shoot should be gently rubbed off and the young plants must be cosseted – no ploughing in their vicinity, only a prong hoe used by hand to control weed growth in the exposed soil around the vines.

> Wilkinson preferred to plant new vines in June: 'one good watering may mean the difference between a good strike and a complete failure'.

In the first year of growth, vine pruners should leave one tender green shoot on each plant, and retain two buds on that shoot. By the second year, the vine should reach the espalier wire, and snipping the plant 3 inches below the first trellis wire would make the central, single vine trunk the 'standard'. This vine trunk was to be tied to the wire, and then two or three new growth canes left on each side. In year three, one of the vine canes on each side of the plant – located closest to the wire – should be selected to form the 'arms, which should be 18 inches to 21 inches long from the standard. These arms should be twisted round the wires and tied there'.

In reaching the fourth and following years of vine growth, Wilkinson concluded, either new arms (from canes of the previous year) should be tied down to the wire, or 'put down' – or the pruners could 'spur prune': leaving five or six vine shoots on each vine arm. What of the wine from these carefully husbanded vines?

Wilkinson described to his listeners that the quantity of fruit sugar naturally occurring in different grape varieties – still measured with Keene's saccharometer – defined qualities of taste that governed labelling for style. For example, said Wilkinson, 'the hermitage or Black Shiraz grape picked with 18 per cent to 21 per cent of sugar' made 'a light claret', and the same grape variety picked at 22 per cent to 25 per cent of sugar resulted in a 'Burgundy'. When the Brix level topped 26 per cent and beyond, the wine should be fortified and branded as port. (Unfortunately, he does not give a description of how he classed his white wines according to grapes used.)

Farming systems – while deeply influenced by the contingencies of local environmental conditions – do not occur in a vacuum when it comes to deciding how to respond to nature or technology, let alone markets or fashion for wine. More than any other generation of the Hunter wine community, the forgotten generation was making decisions informed by the decisions of others – at the layers of the state and the nation, and internationally. The multifactorial nature of his winegrowing can be emphasised by focusing on the moment in which his calloused hand reached out to remove a shoot from a single vine (or to prune and tie the vines); and then widening the view to encompass his brother Garth and hired labourers also performing this meticulous procedure on perhaps 5000, 10 000 or 20 000 grapevines, often for weeks on end; and then envisaging the same labour occurring on neighbouring vineyards – in an instance of simply *one* of the fundamental bases on which rest those higher scales of markets and the grand networks of highways of maritime trade. Wilkinson won his knowledge striding in work boots across his farm and neighbourhood – marking the barest filigree of webs of connection compared with activities of larger, wider worlds. By both engaging in and *sharing* his experience within a farming system, Wilkinson provided connection and reconnection within the circuitries of his neighbourhood of winegrowers and other members of the farming district who collaborated to maintain the contribution of their locale to the wider Australian wine industry.

The passing of old traditions

LIKE SEVERAL PROPERTIES IN THE PATERSON DISTRICT, CAERGWRLE IS nestled into the landscape, isolated within closer, lusher country than west of the Hunter. At Caergwrle – a mixed farm like most in the region – grapes were grown on the river flats, close to the homestead. The 50 foot long shed deployed as a cellar was concreted with a sloping floor some time early during the forgotten generation, and wine from as many as four other properties was stored there.

Richard Boydell recollected that there were two wine presses, one for white grapes and one for red, and at the end of each day all of the winery equipment was scalded in enormous cast-iron coppers heated over a wood-fire. 'Sunday was usually the day a lot of locals came to buy their demijohns of wine', he said. 'Some small bulk lots were sold to Mrs Bird's Wine Shop', a well-known local landmark, 'however the main bulk was pumped into casks to go to McWilliam's Wine through Maurice O'Shea' until grape growing ceased in 1933.[57]

By the end of the forgotten generation the acreage extent of Hunter vineyards, so widely visible in 1901, was clearly diminishing. Yet members of the Hunter wine community clustered around Pokolbin enjoyed more suitable environmental conditions compared with some other parts of the region, and were able to bear the forces of external change through the high-tensile nature of community connections, manifest in bonds of trust and reliance, financial prudence, and experience – to maintain a locale.

Prophecies of doom

IN 1929, WITH AUSTRALIAN WINEGROWING IN A STATE OF COLLAPSE DUE to downy mildew in some districts and the Doradillo bonus in others (and despite the postwar tightening of temperance values and policies), Federal Parliament called for an inquiry into the need to rehabilitate the national industry. The inquiry marked the simultaneous end of the first generation of a national wine industry and the fourth generation of Hunter winegrowing. Nationally, 1929 also saw the massive economic bust in international finance that caused the Great Depression. Given that it was a primarily urban nation by this time, most historians have focused on the devastating social impact of unemployment in cities, where most Australians lived. One of the effects of the Depression at Pokolbin and surrounds was that for a few years hundreds of unemployed coal miners were employed at grape harvest, instead of the women and children of the district who had long increased the workforce for this seasonal labour.

The 1931 Report on the Wine Industry of Australia gave a snapshot of the Hunter as 'eminently suitable to the production of high quality dry wines' – with sandy, volcanic and clay loam soils – renowned for Hock, claret, Chablis and Burgundy styles from low grape yields of less than 1 ton per acre, on average. 'In almost every season the yield is adversely affected by hail, frost, or flood', and seasonal temperature ranges were less severe than in other wine regions.[58] Rainfall averaged 28.40 inches per year. The principal grape varieties were Semillon, Shiraz and 'White Hermitage', Blanquette, and some Madeira and Pinot (perhaps Pinot Noir, but more likely Chardonnay). In 1930, there were close to 2000 acres under vines, among 49 individual vineyard owners. In this district, more than any other Australian wine region, the grape growers made their own wine.

But as demand for dry wines remained scant, the investigators concluded that *'it appears doubtful whether this area can continue in production'* (our emphasis).[59]

Following on from this, a report soon appeared in the Sydney press, stating that the Hunter winegrowing community numbered closer to 100 growers in the region, and claiming that if they faced a bleak future it was because the federal government supported the 'sweet wine trade' over the production of dry wines (the Hocks and clarets) that had earned the Hunter its fine reputation.[60]

The notion of the Hunter as unsuitable for winegrowing depended on an export focus on large-scale production of fortified wines and brandy that unfairly persisted even into the late 20th century.

To encourage Australian wine trade, the federal government passed the *Wine Overseas Marketing Act* of 1929–30, which led to the establishment of the Wine Overseas Marketing Board in 1931. The first incarnation of this board did not include a Hunter representative.[61] Winegrowers in the Hunter did not benefit from the board's control of export prices or bounties in support of grape growing in other districts. The Hunter's lighter style wines did not travel well, and without refrigeration were not suitable – as were sweet wines – for export to the northern hemisphere. Moreover, Hunter wines could not compete with similar French wines on price in the British wine market because of the higher cost of labour in Australia than in France.

In 1938 the national wine story – from the Overseas Marketing Board – sidestepped the Hunter's role as a winegrowing community of continuing heritage.[62]

The art of resilience

HITTING BACK AGAINST THE PROPHECIES OF THE HUNTER REGION'S demise, competition from other Australian wine regions, and the lack of federal government interest in supporting (light, dry) table-wine production, the story that Hunter winegrowers told in 1940 was that they were masters of the art of winemaking. Indeed, that 'the delicacy of flavour and bouquet are the envy of winemakers' from other parts of Australia.[63] John Younie Tulloch appeared as the central figure in the story for a regional newspaper report for the vintage of 1940, only months before he passed away in August of that year. The Hunter's 'soil, seasons and locality are basic necessities, but there is still another factor in the production of quality wines. The winemaker must know his business' – and Tulloch's contribution spanned nearly half a century. 'It is noteworthy', the report continued, 'that soil and season determine the variety of the wine. At Pokolbin, for instance, where Reisling [sic] grapes will produce hock, and Pinot grapes Chablis, the same vines planted elsewhere might produce wine of an entirely different character. There are even differences in type from the same vines in different parts of the one vineyard'.

Only one in 20 people preferred the type of dry wines produced in the Hunter. The report proceeded to explain the difference between the production of Hocks and clarets, and sweet wines, as a form of wine education that was before its time. Tulloch expressed his hope that (since Hunter wines were not aiming to export) the state government would introduce legislation 'to allow the people to get light wines served with their meals at restaurants during the midday meal and between 6 and 8pm', which had been the custom in Victoria and South Australia.[64]

Cellarmasters during the temperance era

IN 1901, WALTER PETHERBRIDGE OF PETHERBRIDGE'S WINE, SPIRIT AND Provision Store in Maitland purchased wine in bulk, matured it in barrels in his extensive cellars, bottled the wines ready for sale 'with the latest and most improved machinery', and labelled them with his Sun brand – for a product made from 'pure juice of the grape' and quality 'first class'.[65] From 1916, licensing laws focused on limiting drinking in pubs, clubs and wine

Lindeman's retail wine cellar, Queen Victoria Building, c1920. Sydney Reference Collection NSCA CRS 51/2333 City of Sydney Archives.

McWilliam's Wines; Mount Pleasant, Darby Street Newcastle, c1946. State Library of NSW, Call no. Home and Away – 32192.

shops. No such restrictions applied to buying wine from Petherbridge's, or from Lindeman's Cawarra Cellars in George Street, Sydney, or later from Johnnie Walker's Rhine Castle Cellars in Sydney and, from the 1940s, the art deco beauty of McWilliam's Mount Pleasant wine saloon in Darby Street, Newcastle, which stocked wines produced by O'Shea.

According to Australia's maestro of wine writing, James Halliday, when his father returned from studying at the Royal College of Physicians in England in the 1940s, Halliday senior began to stock his cellar from Lindeman's in Sydney. The cellars were managed at this time by Leo Buring, who had been appointed by receivers after the company's insolvency. Halliday's father became an 'important private customer' of Lindeman's Ben Ean, Coolalta and Cawarra brands, buying only from Buring, and 'reserving particular scorn for what he termed "South Australian jam" [fuller bodied wines]'.[66]

In 1947 shorter licensing hours were again the subject of a compulsory referendum. This time 10 am opening of licensed premises enacted during the war remained fixed, but voters were asked their views on retaining six o'clock closing. The choices for closing times were 6 pm, 10 pm or 11 pm.[67] In this referendum, an overwhelming majority of voters still preferred 6 pm closing – with a ratio of three to one against the later times among voters in Newcastle, Maitland and Cessnock. Temperance activists were elated by the result, while liquor licencees were resigned to continuing limitations on their trade.[68]

As an Arts/Law student at the University of Sydney in the mid-1950s, Halliday belonged to the College Wine Club at St Paul's College, which ordered Penfold's, Tulloch's and Elliott's wines from Johnnie Walker.

Halliday travelled with his father to the Hunter to meet with Lindeman's Hunter manager Ray Kidd and Bordeaux-trained winemaker Gerry Sissingh, as a private customer. Halliday also recalls travelling in the 1950s to the Hunter from Sydney to:

> feast on massive T-bone steaks cooked on a barbecue at Tulloch's, seated at a rough-hewn table on the earth floor of the winery, with 'Riesling' followed by Dry Red and Private Bin Dry Red, by sheer luck often the famous 1954 Private Bin, which caused a furore by winning trophies at the Royal Sydney Wine Show for Best Claret and Best Burgundy in the same year.[69]

(Clarets and Burgundy were, of course, both made from Shiraz grapes.)

Taking a broad view of how to buy Hunter wines in the 1950s, Halliday remembers that O'Shea's wines were marketed by McWilliam's (at outlets such as the Newcastle wine saloon) and only Hector Tulloch and his family sold wines to the public, from the Glen Elgin cellar door. Bob Elliott's wines were sold at a shop in Cessnock (Elliott, a second generation producer at Oakvale, also sold to O'Shea). All other producers in the region sold their wines in bulk (barrels) to other winemakers, or to wine merchants for bottling, labelling and retail.

Perc McGuigan, Penfold's Hunter Valley

BORN IN 1913, NEAR MAITLAND, PERC MCGUIGAN SPENT MOST OF HIS early life at Rothbury, the small district neighbouring Pokolbin. 'And when we moved to Rothbury', remembers McGuigan, 'of course we had a little vineyard, only about eight or ten rows, but there was a fifteen-acre vineyard some two or three miles away which my father helped the owners to work'. McGuigan woke early every day to bring in the dairy cows for milking, and on school days rode his horse to Mistletoe Public. From a young age, he worked in vineyards during the vintage. And, as a matter of fact:

> I used to make cartridges and shoot birds in the years 1921 to '23. When I was only eight to ten years old I used to make the cartridges and shoot the birds at the HVD vineyard – the Hunter Valley distillery vineyard – at Pokolbin, which as history turned out, I was given the management of that particular vineyard in 1941.

McGuigan's career with Penfold's did not progress directly from bird scarer to manager's chair. His father died in 1927, and four years later – as McGuigan prepared to sit his Leaving Certificate (the final year of high school, equivalent to today's Year 12) – he received a job offer to clerk at the Branxton butter factory, which he accepted, to assist his family. Over ensuing years, he studied by correspondence to gain qualifications in dairy chemistry and dairy product grading. But the decision to accept the position at Penfold's Dalwood headquarters in 1941 catapulted McGuigan into the wine industry in the midst of the era of the stoic generation.

McGuigan had been at Dalwood for only five or six weeks when Japanese forces attacked the US naval base at Pearl Harbor in Hawaii, spurring America's entry into the Pacific theatre of World War II. 'We had fourteen employees' at Dalwood, 'and seven of those were taken' to serve in the oncoming war. 'And I can assure you', recalls McGuigan:

> that the next four or five years were pretty desperate because it was difficult to get anybody to help in the vineyards. And because of the

Perc McGuigan (front), with wife Sylvia and sons Ross and Brian, c1950. Imprint by Peter Stoop. McGuigan Family Collection.

fact that a lot of people were taken from the vineyards. And it was only the fact that the people who stayed on were able to maintain those vineyards sufficiently so that when the men came back from the fighting forces we were able to continue on from there.

Retired labourers were coaxed to replace lost hands at Penfold's vineyards, along with men who, for whatever reason, could not go to war.

During and after the war, Penfold's continued to make light (or dry) whites and reds, but most wines were fortified styles branded as port, sherry or Muscat. Once produced, the wines were transported directly to Sydney headquarters – or elsewhere within the web of connections that Penfold's company had established across the south-eastern quarter of Australia, linking the hub of Adelaide with the satellites of the Hunter (and the Upper Hunter), Minchinbury in Sydney (renowned for sparkling wines), and Griffith in the Riverina of New South Wales. By 1950, McGuigan managed four Hunter vineyards totalling over 200 acres – with his largest vintage being 65 000 gallons, which was considerable in its day.

Electricity did not become available in Branxton until 1946. 'So you might wonder', asked Perc, 'how we made wine at all when we weren't able to control the rate of fermentation? Well, we used to get eight and ten tons of ice', from Maitland.

> And of course, in those days ... we used to have at least eight or ten consecutive days where the temperature reached 100 degrees Fahrenheit ... and the nights never got below 85 degrees Fahrenheit, which would mean that when you went to pick the grapes the next morning they were ready to be picked but when the grapes came in they were testing 85 degrees, and there's no way in the world that you can make a good wine – a top wine – if the temperature gets up above 90 degrees.

The solution?

> Well, what we used to have to do if the grapes came in at that temperature – we would start, of course, at daylight but by eleven o'clock we'd probably have to get the pickers to go home and pick them up at daylight next morning, and it was hoped that during the evening a southerly change would come and it would drop the temperature. Because, even using ice, it was difficult to drop a tank of juice down – if you used two or three tons of ice, it was difficult to drop a tank of juice down any more than five or six degrees Fahrenheit. And within an hour, it was back up to that temperature again.

With the advent of refrigeration powered by electricity, said Perc:

> we just smiled at the weather. And we liked it to get very hot, and the nights to be hot, because immediately the grapes came in we could drop the temperature down to about 50 degrees Fahrenheit. Let it ferment away for a fortnight or whatever it was. The juice used to ferment out in three days. If we had no way of stopping it from racing away it used to ferment out in three days. We still made fairly good wine but how we did it is quite a problem. But we did do it.

Hunter wines, changing taste

AT DIFFERENT TIMES IN 1949 AND 1950, MAX DUPAIN – ONE OF Australia's most brilliant photographers – visited Pokolbin, depicting McWilliam's Mount Pleasant and Maurice O'Shea, and Hector Tulloch at Glen Elgin. These images show the harvest taking place at Mount

Pokolbin, c1949. Photograph by Max Dupain. Tulloch Family Collection, courtesy of J.Y. Tulloch.

Pokolbin, c1949. Photograph by Max Dupain. Tulloch Family Collection, courtesy of J.Y. Tulloch.

Pleasant, with men and women among the vines, and barrels of freshly harvested grapes carted by horse-drawn dray to the winery – where O'Shea used a wooden basket press to make the must that was fermented in large, square, open concrete fermentation tanks. Men are cleaning away the cake of pressed skins and stalks from the press, and preparing large storage barrels.

Tulloch is photographed amid his bare vines, during winter, and in his wine cellar.

From as many as 100 grape growers that the wine community had counted in 1931, as few as ten remained in 1954, and most of these concerns were owned by large firms based in the region, or beyond.

In 1954, Audrey Wilkinson noted that of the first 22 members of the PDVGA in 1901 'only four names remained connected with the wine industry'. He presumably meant himself, Tulloch, Tyrrell and Drayton – and he blamed the loss on modern methods of mechanisation that increased the scale of winegrowing elsewhere.[70]

Although there were changes that were unwelcome to Wilkinson, some other alterations assured the Hunter's continuing competitiveness in an evolving consumer market. No longer in the 1950s did Hunter wines have the literal aromatic astringency of gum trees that Bordeaux judges perceived in 1882. Still, the habits of Hunter wine men – growers, makers, traders – were governed largely by travel on dusty roads, draughthorses to draw ploughs and drays, corrugated iron sheds for wineries – with rammed earth floors and open concrete vats. Amid the remnant blue-green bushland on the Brokenback Range that dominated Pokolbin, and throughout the softly contoured landscape of the other dispersed districts of the region, figures such as Maurice O'Shea and Hector Tulloch bought grapes and wine from many vineyards, blended experimentally to suit the variations of the market, and demonstrated an evolving mastery of vinegrowing and winemaking. After the arrival of electricity services, winemakers began to adopt the newer technologies described by McGuigan – refrigeration, larger fermentation capacities, and machines for separating wine from the debris of the crushed fruit – to exert greater control over styles and flavours.

Examining barrels, Mount Pleasant winery, New South Wales, c1950. Photograph by Max Dupain. National Library of Australia, Call no. PIC/8729/2 LOC Album 757.

At the end of the generation of the stoics, during which hardship combined with curiosity became the engine for ingenuity, some winemakers possessed the skills to orchestrate wines more broadly than before. This handful of men were sometimes working in small firms, but were always tied to the wider winegrowing community. Others were employed in large multiregional firms.

Fine wines were the handiwork of winemakers who had gained an understanding of the quality of the grapes they grew or purchased. This is where winemakers in larger wine firms in the Hunter were critical contributors. By casting wide nets across the Hunter (and other regions) to obtain grapes and continue operating, they achieved a depth of understanding of the possibility of using grapes from certain vineyards – combined with technologies to hand – to craft wines that would prove to be long-lived and of great distinction.

Each of these winemakers were now standing on the shoulders of previous generations as they developed their skill though trial and error – with an overarching view of grape varieties and their characteristics, from a range of vineyards. These winemakers knew which grapes, from where, they required for certain purposes; how long to ferment the must; which barrels were best; how to regulate temperatures; clarify wines; and so on. Most significantly the stoics were no longer imitating wine styles from other places – as had the imaginers and experimenters. In the years between 1930 and the mid-1950s, the few remaining Hunter winemakers – in continual conversation with grape growers and other winemakers – divined how to both make wines of Hunter locale and draw together the strengths of taste from grapes of different regions, which proved to be a critical skill in the incoming era of a changing taste for wine.

Pokolbin, c1949. Photograph by Max Dupain. Tulloch Family Collection, courtesy of J.Y. Tulloch.

AMID TURMOIL AND TEMPERANCE 229

Chapter 8

Civilised drinking in a wine nation: the Renaissance generation, 1955–83

IN 1955, MANY MONTHS OF HEAVY RAINFALL FLOODED THE HUNTER River system, causing the loss of dozens of human lives and thousands of livestock, and damaging thousands of homes and farms. As the waters receded, Karl Stockhausen, a 25-year-old industrial company clerk from Hamburg in Germany, arrived in Australia seeking a new life distant from his postwar homeland. From Sydney, Stockhausen travelled by train to Greta Migrant Camp near Branxton in the Lower Hunter, making a journey from an old (in fact, flood-prone) European city, halfway across the globe and deep into a bucolic countryside. The camp at Greta, housed in a former military barracks, was one of the nation's largest sites in the federal Department of Immigration's postwar immigration scheme, providing temporary housing and cultural education for more than 100 000 people between 1949 and 1960. The small Hunter winegrowing community hired seasonal labourers from the camp.

As Stockhausen's train steamed through Maitland soon after the floods, he was alarmed by the sight of mud 2 feet deep in public parks, and grass snagged on telegraph wires 20 feet above the ground.¹ Despite these dispiriting impressions, more broadly Stockhausen's arrival accorded with an upward swing in the national mood after two generations of the devastation of war and economic hardship in Australia. From the mid-1950s, a dawning restiveness, an eagerness for greater conviviality, among women as well as men, buoyed a relaxation of restrictions on drinking rights and the emergence of new drinking rites.[2] Environmental historian Eric Rolls – a great lover of Australian wine – observed of the 1950s, 'men drank beer or rum, the ladies drank shandies, or perhaps sweet sherry. For a man to drink table wine was abnormal, unhealthy, and unsexing; he was generally suspect'.[3] Postwar immigrant men from wine-drinking countries in Europe in the 1950s were criticised by Australian men for preferring homeland habits of wine drinking to beer.[4] But change was coming.

From the 1960s, the protest movements calling for an end to the Vietnam War, and for civil rights for women and Aboriginal people, were part of a wider desire for inclusive and peaceful attitudes to daily life. This new ethos opened the way for change in wine-drinking habits, which were in turn encouraged by the marketing-savvy wine industry.[5]

Economic confidence in the wake of World War II, combined with a hopeful social climate, propelled the expansion of vineyards and wineries across Australia's wine states, the development of new wines, and the growth of wine communities, including in the Hunter.

For the Hunter wine community, the years from 1955 to the 1980s were a Renaissance – a shaking out of older production practices that limited the characteristics of products available to a generation thirsty for wine, and a shaking up of the skills required to maintain drinkers' interest amid changing fashions.

Civilised drinking

THE EARLY CLOSING OF PUBS HAD GIVEN RISE TO THE INFAMOUS 'SIX o'clock swill': men binge drinking from the time they finished work in the afternoon until the pub's doors were shut at 6 pm.[6] This was only fun for the participants, and by 1950 Australian public drinking habits meant irreverent, boisterous, (usually) working class men drinking beer in pubs. In cities and in towns, there was barely any hospitable nightlife for men eschewing the 'swill', or for women – let alone lunches served with wine in restaurants or cafes. In 1951, the New South Wales government established the Royal Commission on Liquor Laws. The purpose of the commission was to inquire into concerns over the business connections between brewing companies and pubs, and the opportunity arose to also hear evidence about the lack of alternatives to the pubs for public drinking. Presided over by Justice Allan Maxwell, the commission sat from 1952 for 140 days, and examined more than 400 witnesses.[7] Maxwell found clear evidence of an appetite for the relaxation of access to alcohol, within the state legislature – representing the broader community – as well as in the liquor trade. Some resistance was offered by the religious leadership of the state, and by the Temperance Alliance. The commissioner reminded these few remaining agitators for blanket prohibition of all alcohol, however, that 'facilities for liquor in its various forms is something which is accepted by and part of, the law of the land'.[8] Australians would continue to drink alcohol, stated Maxwell, and measures ought to be introduced to encourage drinking *without* drunkenness.

A subsequent 1955 referendum on the extension of trading hours for licensed premises from 6 pm to 10 pm received only narrow voter support – yet a corner had been turned, away from the six o'clock swill and the path of prohibition of the past 40 years.

The Liquor Commission debated the notion of grocery shops being allowed licences to sell single bottles of wine, and whether to continue to limit the sale of wine in less than 2 gallon quantities to wine shops (many of which Maxwell observed were of evil repute). While bottle shops for beer, wine and spirits were not yet sanctioned, these innovations were on the near horizon.

Branches of the wider system of the wine industry were represented at the Liquor Commission by the Wine and Spirit Merchants' Association (the main sellers of imported liquors), the Australian Wine Producers' Association of NSW, and (jointly) Penfold's Pty Ltd, McWilliam's Wines Pty Ltd, and Caldwell's Wines Ltd (a company based in the Riverina). These three companies owned most wine outlets with proprietary ties. Other groups tied to winegrowing gave evidence. Maxwell remarked that for groups with purportedly shared interests, wine bodies were clearly divided. But this is not surprising. Wine and spirit merchants were in competition with wine companies large enough to be vertically integrated

Gerts from Greta Migrant Centre picking grapes (probably Glen Elgin), 1959. National Archives Australia A1500/K4194.

(where single companies produced grapes, made wine and distributed their products), and the large wine companies – although often grape growers – were also grape buyers, governing prices received by producers. The circuitries of the wine industry, as a system, did not always hum in harmony.

Ultimately, some widening interest in conviviality combined with civility in drinking called for wines with a lower degree of alcohol than fortified styles; the sort of wines where a single glass or two could be enjoyed with a meal to lighten the mood and lubricate conversation: Australian Burgundies, Hock, Chablis and claret.

The Hunter's time had come.

In 1961 it became apparent that Australian wine drinkers were beginning to reject fortified wines in favour of table wines. As early as 1963, wines were advertised in the pages of the *Australian Women's Weekly* – the nation's most influential magazine for women – and campaigns targeting women increased from the late 1960s. This was a dramatic shift from the social expectations of the 1950s, whereby women were considered more respectable if they avoided drinking altogether. Postwar immigration also brought new attitudes to eating and drinking, which arrived with European men and women from wine-drinking countries – some of whom joined the wine industry.[9]

Making Australia a wine-drinking nation

TAKING A LONG VIEW OF CHANGES IN DRINKING CULTURE UP UNTIL 1983: by the end of the 1970s Australians were the highest-volume consumers of white wine in the English-speaking world.[10] This growth had been largely driven by women's embrace of alcohol and a new middle class of Australian men. In the 1970s, the working class basis of Australian society that had existed since the convict origins of the early colonies (alongside a relatively small professional class and few wealthy citizens) underwent a dramatic change. New investment in industrial-scale, mechanised forms of agriculture, mining and manufacture led to a marked reduction in working class jobs for men. Democratised access to tertiary education for the professions – with the federal government funding university tuition fees

– and a more managerial workforce, gave rise to more white-collar employment for men, and for women. Men wearing blue singlets as a symbol of working class Australia began to disappear, and the new (mainly urban) middle class sought new styles of eating and drinking in restaurants and cafes. By the end of the 1970s, wine was no longer unmanly nor unAustralian. Thanks largely to the accessibility of taste and price of wine styles created since the 1960s – several of which had some connection to the Hunter, including Lindeman's Ben Ean Moselle and Wyndham Estate TR2 – Australia had become a wine-drinking country. This also spurred interest in locale-focused wines from the region, as discussed below.

For the Hunter wine community, the postwar Renaissance for the industry grew from, and further strengthened, ties to wider social and cultural worlds – connoisseurs, non-connoisseurs, tourists, writers, celebrities, corporations. After the lowest point of wine grape plantings in 1959 (estimated by James Halliday to have been 500 acres), new vineyards were created from the 1960s and winegrowing began to gather and regather pace, achieving a rate of change that rapidly outpaced the same kind of development in earlier boom periods. Traditions of multiregional connections remained, and Hunter wines in this era existed across a spectrum of products, from those made from one grape variety grown on a small defined patch of vineyard, to wines named for Hunter places and produced from grapes grown elsewhere. The Hunter wine community evolved into a more complex multilayered system of cultivation, manufacture, distribution, promotion, trendsetting, troubleshooting, paying taxes, hiring employees, orchestrating diversity in investments for resilience, and training, retraining and training others.

Because the Hunter wine community comprised deep, local, family traditions as well as ties through national companies to the broader scale of wine business – and because the wine region is close to Sydney – it played an integral role in the new wine era, with several new characters entering the story.

HUNTER RIVER VALLEY AT VINTAGE TI[ME]

By BETTY BEST, staff repor[ter]

For more than 100 years [the] Hunter River Valley, N.S[.W.,] has produced some of [Aus]tralia's finest vintage[s.]

As the hot sun of an Indian summer shone on the vine-covered slopes of the Hun[ter] Valley this season, pickers gathered the re[cord] what should have been a record harvest, b[ut] will produce only about 180,000 gallons of [wine.]

Heavy rains and, in some cases, hailstorms r[obbed] growers of the bumper crops they were expectin[g.]

Red or white, light or heavy, under well-kn[own] labels or in private bins, all the grapes of the [Hunter] Valley are made into dry table wines—the wi[nes of the] connoisseur.

Unlike French wines, most of these are b[lended to] produce a permanent type—year in, year out[. Most] Australians like to buy wine under a regular l[abel, such] as claret or hock, or any well-known type, rath[er than a] yearly vintage from a particular district.

That is why the big firms employ winema[kers who] travel from cellar to cellar at vintage time ble[nding the] fermenting grape juice into standard types that c[onform] as closely as possible with the wines of previous [years.]

Some vintners, however, believe that Austral[ian wines] in their pure vintage form compare favorably wi[th those of] the world.

Hector Tulloch, who inherited "Glen El[gin" and] "Fordwich" vineyards from his father, holds thi[s view.]

"My brothers and I have lived, dreamed, [and been] taught to appreciate wine for as long as any o[f us can] remember," he said.

"I put down my first vintage at the age of fo[urteen."]

"I had to pick, crush, and ferment my own gra[pes, using] my own judgment in the task," he said. "Then [came the] exciting moment when it was ready to cask. [I made my] own 'casks,' too—a neat row of tightly sealed trea[cle tins."]

Now Mr. Tulloch's 320 acres of vines contrib[ute their] flavor to the blended wines of several big firms.

Dozens of private customers also send orders [for their] supplies to his cellar at Pokolbin, and he is on[e of the] chosen few growers who supply yearly vintages to [the 400] connoisseurs who have banded together in the [Australian] Wine Consumers' Co-operative Society.

Just across the hills from "Glen Elgin" lies the [vine]yard of "Mount Pleasant," where vines planted m[any] years ago are still producing some of the best wi[nes of the] Valley.

"Mount Pleasant" is managed by Maurice O['Shea, a] true wine lover, who studied wine-making in F[rance for] six years.

Mr. O'Shea has been bottling and dating his [vin]tages for 20 years. European connoisseurs have [compli]mented him on his wines.

VINE-COVERED SLOPES of "Ben Ean" vineyard at Pokolbin, N.S.W., are alive with pickers at vintage time. The grapes are put into ordinary buckets and carried to a tractor which follows pickers from point to point. Rate of pay is threepence a bucket. This summer's picking record at "Ben Ean" was 188 buckets gathered in a day by a woman.

PRIVATE BIN vintages, bottled for sale, line the walls of Hector Tulloch's cellars, left. Mr. Tulloch is one of the few Australian vintners who believe in bottling the season's vintage rather than blending it to conform to popular types.

RED WINE gets its color from [the] skins pushed to the bottom [of fer]menting vats by boards. H[ector] Russell, 16, and foreman E[...] remove boards which had been [...]

'EADS of freshly picked grapes are poured into a vat before pressing 'Pleasant" vineyard. The manager, Maurice O'Shea (extreme right), weights and Noel Martin tips the hogshead towards chief cellarman Ray Dunnercliff, who is waiting to rake in the juice-laden bunches.

EXPERT PICKER Mrs. H. A. Pincher, 67, of Mount View, above, has been picking grapes for 59 years. Pictures taken by staff photographer Clive Thompson.

EUROPEAN OAK CASKS, which each hold 1000 gallons, are scrubbed out by hand before being filled. Below, Les Stewart, 23, of "Ben Ean," squeezes into one sideways.

STRAINING on the handle of a wooden press, William Pincher and Noel Martin give a final turn to extract the last of the juice which flows through loose slats into the circular tray underneath and is pumped straight into the fermenting vats.

TALLY CLERK A. J. Andrews, 28, of Lonedale, near Pokolbin, empties buckets of grapes into waiting hogsheads and notes the score of each picker. In the height of the season pickers work from daylight to dusk to get the grapes in while fine weather holds.

Hunter River Valley at vintage time, *Australian Women's Weekly*, 6 May 1953.

Karl Stockhausen, Lindeman's winemaker, Pokolbin

WITHIN A WEEK OF ARRIVING AT GRETA MIGRANT CAMP, STOCKHAUSEN was hired to chip weeds (that is, to use a hand hoe to remove plants from between the vines, where a plough could not reach) in Lindeman's vineyard – blistering work, as he recalled! Stockhausen recovered, however, and continued to be employed by Lindeman's (a public company since 1953, at this time, with Ben Ean managed by Hans Mollenhauer, also from Germany). After a few years, Stockhausen's skills as a clerk gained him a role in Lindeman's head office in Sydney – which he remembers as a staff of about eight people. He undertook some management study and would at times return to Lindeman's at Pokolbin, for the vintage, or when Mollenhauer took leave. In these years, Lindeman's manager Ray Kidd (dux of his graduating year of winemaking at South Australia's Roseworthy College) mentored Stockhausen. When Kidd could see that the 1960 grape harvest was so small it might be unprofitable, he left Stockhausen alone to experiment with the vintage, allowing the young German a valuable opportunity to test his winemaking instincts.

Karl Stockhausen at Lindeman's Ben Ean, 1970. Department of Immigration, Multicultural and Indigenous Affairs, National Archives of Australia A1211, 1/1970/16/377.

'So, we did the best we could, and picked the grapes', recalled Stockhausen. 'And it so happened that we turned out a white wine that was absolutely fabulous. 1960 vintage bin number 1616, I shall never forget it', he said, referring to the distinctive practice of 'bin' labelling. (Tyrrell's instead differentiates fine wines by vat numbers, but the concept is the same – a new way of describing new wines.)

'We had, in our winery, a pressure fermentation tank', Stockhausen explained.

Now this was an innovation that was taken from Europe, and the first ones to use it extensively were Orlando in the Barossa Valley

... This was the time of pearl wines – Barossa Pearl – so they had a number of pressure vessels and they also used them for fermentation. A German technique of fermenting grape juice into table wine ... And there was some promise of making better wines, richer wines, fruitier wines, by that method. And therefore, Lindeman's had gone by and had one of those tanks built, and I used it in 1960 and made this fabulous – Semillon. In fact, it wasn't a straight Semillon. It was Semillon that had a bit of Traminer in it, a little bit of Verdelho in it. So to speak, anything that we could rescue from that vintage went into the tank and made this fabulous wine. In those days the wine shows in Australia didn't have gold medals, silver medals, bronze medals. It was a matter of first, second and third prize. And then you got a few recommended or highly recommended. And this wine, wherever it was entered, and whichever class it went into, whether it was the so-called Riesling class, or Hock class, the 8 White Burgundy, Chablis, whatever those classes were, wherever it went, it won first prize. So that was the first success.[11]

Stockhausen's knowledge as a winemaker arose from a combination of mentorship, experience and self-education. He ordered books from Germany on winemaking 'because you couldn't buy any books here that would teach you anything about winemaking at the time' – and these books are still on his shelves in his office at his home near Pokolbin. The instruction manuals were:

very simple, but nevertheless particular about what you do and what you don't do, how to make good wine. And it also, of all other things, was a reference and particular instructions on how to use a pressure fermentation tank, which pressure fermentation tanks didn't exist; all we had basically in the Hunter at that time, and that applies to all wineries, concrete fermentation vats for fermenting juice into wine, and wooden casks; no stainless steel. Electricity had only been put on in 1955, and so that was only then just on, and refrigeration didn't exist. So, it was all very very simple and plain.[12]

The world's finest connoisseur visits the Hunter

THE SAME YEAR AS STOCKHAUSEN ARRIVED AT GRETA MIGRANT CAMP, the Institute of Masters of Wine was inaugurated in London. This group formalised the specialisation of wine knowledge, for traders, sommeliers and writers – and offers the wine world's most influential qualification for wine intermediaries, the Masters of Wine (MW, listed after the name of recipients). The inception of the Masters of Wine program signalled a rise in wine connoisseurship that would accompany a concurrent interest in wine more generally in the English-speaking world of the postwar era. In 1966, Frenchman André Simon – the most significant aficionado and writer operating in the historic British wine trade – visited Australia, at the invitation of the Australian Wine Board. Johnnie Walker's Wine and Food Society, the forum in which the Hunter's fine wines were appreciated by Sydneysiders, had been established in imitation of Simon's society for a fine wine fraternity in London.

(During his travels in Australia, Simon found that one of the large wine casks at Lindeman's Corowa property in the Riverina came to this country as war reparations from Germany. He did not say whether similar equipment came to the Hunter.[13] Europe's war-torn wine industry struggled to recover in the second half of the century; and indeed, as Australians turned *towards* wine, wine-producing, wine-drinking countries were turning *away* from their wine traditions, for many reasons.[14])

In the Hunter, Simon found 'many of the lighter and better Australian dry red table wines' with an average alcoholic strength of 13 per cent of alcohol.[15] Penfold's Dalwood was the 'oldest' property, Lindeman's Ben Ean, the 'largest', and McWilliam's Mount Pleasant, 'one of the best known'. Simon also visited Tulloch's at Glen Elgin and Tyrrell's at Ashmans, as well as Elliott's Oakvale and Drayton's Bellevue (and other sites), commenting especially on the quality of Semillon wines (still called Riesling, which Simon advised changing). He thought the Hunter River system had 'not too many friends, but too many tributaries'.¹⁶ The first part of this rather oblique statement in all likelihood refers to evidence in the 1931 report, and to hostility from state government employees and the wider wine industry towards the Hunter, which made the region seemed friendless. The second part of the statement conjures the region's beginnings along the Hunter River and the branches of the Lower Hunter region, and the subsequent contraction of Lower Hunter plantings away from flood-prone areas.

Simon also noticed that Cabernet Sauvignon had been planted at Lake's Folly, Pokolbin, in 1963.

Max Lake, Sydney surgeon, Lake's Folly (Pokolbin)

BORN IN 1924 TO AMERICAN FILM INDUSTRY PARENTS, SYDNEY SURGEON Max Lake not only created Lake's Folly as one of the most lasting, select Australian wine brands (specialising in Cabernet Sauvignon) – his is also an impressive legacy of writing with evocative viscerality on wine and the wine world. In Lake's telling, the Hunter landscape possessed classical grace, and the weather a beloved perversity. 'White woollies gambol over the lion dozing at Brokenback', he wrote, describing clouds over the escarpment that frames the Pokolbin landscape.¹⁷

> Fierce summer breath of the tropics not far to the north; lazy hum of the cicadas; sometimes the tail lash of a Coral Sea cyclone; these scarcely hint of a great wine area. And yet, the wines, like great ones anywhere, are only just made, each vintage; dodging most years, between disasters of one kind or another; to emerge in the glass, as one of the marvellous gifts of capricious nature.

By Lake's own account, his interest in Hunter wine arose from drinking a 1930 Dalwood Cabernet Sauvignon/Petit Verdot.

Lake was at the vanguard of new vineyard owners and winemakers in the Hunter in the early 1960s. Like those among the generation of imaginers and experimenters who had no need of profit from winegrowing (say, Wentworth and Keene), Lake at first continued his medical practice.

Unlike those earlier Hunter winegrowers, Lake eventually resigned from his medical career in favour of winegrowing, when aged in his fifties, to devote himself to his new calling.

For Lake, the white wines of the Lower Hunter were:

> unique in the world, whether made by open fermentation and matured in old oak casks by traditionalists, or gem-polished in steel tanks by trained oenologists. The area bouquet and flavour are perfect foils for the Semillon grape character, with devotees of Traminer, Chardonnay, Blanquette, Marsanne, Aucerot, White Hermitage and others proclaiming the virtues of their selections; picked early so that they are acid and light bodied, or later, so there is a fuller flavoured and softer wine.

'The reds are outstanding in so many ways', Lake continued expansively. 'Here again, a distinct flavour gives the regional Red Hermitage grape an additional interest with superb Cabernet Sauvignon and Pinot Noir wines in many years, influenced as in other great areas, by the climate of the growing season.'

(Many years later Stockhausen would list the vineyards he knew produced the best reds in the region, saying, 'I have no doubt the best red grapes come of heavier soils ... red clay', or as Wilkinson called it, marl.)[18]

Lake went on to say, 'in the Cessnock zone [the growing season] starts in early Spring, August and September of one calendar year, and comes to ripe fruit during February and March of the next. Europeans tend to get confused with our "two year" vintage, because of their grape cycles in the one calendar year'. He recognised that Hunter wines appealed to 'connoisseur and beginner alike'.

Introduction wines

IN THE 1960S, THIRD GENERATION HUNTER WINEGROWER MAX DRAYTON sensed that people were seeking out a 'fancy bit of wine'.[19] By this Drayton meant table wines for polite or refined drinking experiences, rather than fortified styles. But a divide remained to be bridged to create a popular wine culture that would be the bedrock of Australia as a wine-drinking nation, as in European wine countries. To do this, the wine industry began to develop more wines for beginners, as Lake termed them, or introduction wines, as they are also called. These are wines that are affordable for most people, have a predictable taste that may be consumed with or without food (and with almost any type of food), and which do not require specialist knowledge such as whether to drink now or to cellar for later consumption. There are many famous Australian introduction wines – including Barossa Pearl and Porphyry Pearl – and Ben Ean, TR2 and so on. Lindeman's Ben Ean Moselle acquired the name of John Murray Macdonald's Pokolbin farm because it was invented at that site, but was later made largely from Trebbiano grapes grown outside of the Hunter. Naming this wine Moselle carried on the tradition of a European wine lexicon to describe wine style, rather than constituent grape varieties. As with Burgundy in Australia, Moselle did not in fact resemble the wines from that Franco-German region.

Former Lindeman's chief winemaker Philip Laffer described the chance origins of Lindeman's Ben Ean Moselle in the Hunter, and its rapid success:

> The first vintage of Ben Ean Moselle was 1956, which was an accident – a mistake. And to overcome the mistake Ray Kidd decided to call it Moselle. The only name they could think of was Ben Ean, and they sold it, and it worked so well that the exercise was then repeated intentionally, and ultimately became the biggest wine brand in Australia.

The thirst for Ben Ean seemed limitless. 'I think at one stage, at its peak', said Laffer:

> every third bottle of white wine sold in Australia was Ben Ean Moselle. It was quite extraordinary, and the sort of phenomena [sic] that could never ever happen again. I mean, it didn't last for long

but it was an extraordinarily important wine for Lindeman's, but I think it was an extraordinarily important wine for the industry.[20]

Advertisements for Ben Ean in the *Australian Women's Weekly* pitched the wine as 'just right' to drink with any meal (except, we note, breakfast) and any summer foods.[21]

Wyndham Estate co-owner Digby Matheson (in partnership with Brian McGuigan and others) observed a growing trend in the 1960s of young urban office workers engaged in responsible, friendly, public drinking:

> I can remember particularly that lunchtime on a Friday was a big deal, and you'd get twelve or fifteen girls sitting there all having [Hamilton] Ewell or [Lindeman's] Ben Ean [Moselle] or what have you, and you could see that it was a movement where people accepting the new culinary arts from the individuals coming from overseas and trying to find the wines to match these, was giving rise to a big swell of interest in wines.[22]

Women drinking wine at lunchtime in a city cafe was in great contrast to the six o'clock swill of the previous decade.

For men and women coming of age from the late 1960s, wine became the ideal accompaniment to the giddy informality of new sexual freedoms that allowed men and women to engage in the joyous pleasure of new forms of music, parties, film and television.

Visiting the vineyards: tourists and festivals

FROM THE 1960S, CAR OWNERSHIP INCREASED AND OPENED UP NEW FORMS of entertainment for men and women to enjoy together. In 1965, Lower Hunter winegrowers held the first Cessnock Vintage Festival – attracting people who travelled by car and bus to the wine region. Every two years after this, the celebrations grew, along with vine plantings and interest in wine, so that by 1973 the Hunter Vintage Festival incorporated the Upper Hunter vineyards and wineries, which had sprung up in less than a decade.

Lindeman's Ben Ean, Pokolbin, c1970. Photograph by Douglass Baglin. National Archives Australia B942, WINE [13].

(By 1977, Arrowfield at Jerry's Plains between Singleton and Muswellbrook would claim to be the largest vineyard in Australia. Arrowfield has been renamed Hollydene, which was the name of another Upper Hunter winery in the '70s.) Many thousands of tourists flooded into the district from Sydney, Newcastle and further afield. The vintage festivals were a party for the people of the wine region that anyone could join in, and an opportunity to educate a new generation of Australians to become wine lovers.

At cellar doors (the name given to the part of the winery where sales are made) and on wine tours, tourists and winery staff shared views about taste and drinking experiences that allowed members of the Hunter wine industry to tailor their wines more closely to drinkers' preferences. The popularisation of wine inspired other culture makers. Poet Geoff Page lyricised the experience of wine tourism in the late 1960s.

WINE BUYING: POKOLBIN

A strand of sun
at Tulloch's marks
the ends of timber casks.
Four new glasses
stand on a table.
Dollar fifteen
for last year's red.

At Drayton's, the light
cuts down through glass
to stainless steel and concrete.
Cocktail prices
climb the wall.
A thin new white
for seventy-five.

Being of my time, no doubt,
and maybe class,
I stacked three dozen
white on top
of eighteen red
and drove away.[23]

While there do not appear to be any songs about Hunter wine in this era, there is a movie. In 1968, a film crew visited McWilliam's Pyrmont laboratory in Sydney and the Mount Pleasant vineyards and winery at Pokolbin to shoot *Squeeze A Flower* (released in 1970). This madcap comedy of errors, written by American Charles Isaacs, features an Italian monk (played by Walter Chiari) with a secret liqueur recipe who hides his identity as a man of religion from the Italian wine family who employ him. Irish comedian Dave Allen appeared as the son of the wine family. Rowena Wallace starred as the advertising executive attracted to the monk, but unaware of his vows of celibacy. The film title is taken from of the phrase 'squeeze a flower, squeeze a grape', and captures the heady romance between wine drinkers, the wine industry and the advertising industry of the era – with a nod to postwar European migration and themes of unrequited love involving nuns and priests and non-clerics, in an informal era of secular irreverence.

Phil Ryan was a young chemist with McWilliam's at Pyrmont when the film crew for *Squeeze A Flower* took over the laboratory for a week in 1968.

Phil Ryan, McWilliam's Mount Pleasant (Pokolbin)

BORN AT THE END OF WORLD WAR II IN THE NORTH OF ENGLAND, PHIL Ryan immigrated to Australia with his parents and twin brother in 1951, and entered the wine industry after completing a chemistry diploma in Sydney in the early 1960s.[24] In 1965, Ryan secured a position in the McWilliam's Wines laboratory at Pyrmont, also in Sydney – after an unhappy year employed at a paint manufacturing plant, where he found his finely tuned sense of smell assaulted by caustic fumes. (While at Pyrmont, Ryan upgraded his qualifications to a Bachelor of Applied Science.) Enchanted to be surrounded instead by winey aromas, Ryan was also delighted that his managers at McWilliam's expected him to sample wine regularly, to understand the spectrum of wine characteristics, and to experience wines from overseas as well as Australia.

McWilliam's Mount Pleasant winemaker Phil Ryan, c1970s. Ryan Family Collection.

Although the McWilliam's Wines company began at Hanwood near Griffith in the Riverina, the McWilliam family developed close community ties in Pokolbin through ownership of Mount Pleasant. They were, in turn, linked to a wider community in Sydney – the cluster of services required for a large enterprise. During the industry's Renaissance, McWilliam's office and laboratory at Pyrmont operated as a company hub between vineyards and wineries at Hanwood and the Hunter. The Pyrmont headquarters provided a centralised technical base for the company's wine chemists and a Sydney presence that connected the company to metropolitan executives, to Australia's international entrepot for business and trade, and to people seeking wine education.

Once Phil Ryan began to work with McWilliam's, he would be accosted at dinner parties and asked to provide advice to people desperate to understand how to choose and drink wine. 'There was this genuine interest, but people were really struggling', said Ryan. 'They'd hear something that was

55. [Men] of the Drayton family, Bellevue, Hunter Valley; [their] vineyard almost a century old, 19 April 1966. Photograph from Tourism Australia. State Library of NSW, PXA 907 Box 5.

incorrect, and you'd have to steer them back.' McWilliam's changed their marketing strategy to suggest that people drink wines with food. And then, quite quickly, 'there were a lot of wine clubs and wine appreciation clubs', recalls Ryan, 'and wine lectures. It just went on. Every newspaper had a wine writer. There was a wine column in every magazine. So, there was this great thirst for knowledge ... it was good to be in that engine room', with the explosion of interest in wine selling and drinking.

As wine consumption rose, new methods were applied to maintain a consonance with popular taste. 'The industry moved towards gentler processes', explained Ryan. These technologies included stainless-steel fermentation and storage tanks, combined with refrigeration, airbag presses (which crushed grapes to release juice, without leaving such great quantities of grape matter in the juice that it would oxidise and impact on fermentation), centrifuges and vacuum equipment to remove lees, and new ways of finely balancing the addition of yeasts (cultured from other grapes) to the ferment.

Ryan first drove up from Sydney to the Hunter in 1967, to observe the vintage at Mount Pleasant. Most memorable, however, was the 1971 vintage in the Hunter – a 'debacle', as Ryan recalled.

> It showed everything that can go wrong in a vintage. There was no sun from about Christmas Day that year, right through to March, and it just rained continually, and it was bitterly cold. But it was still fascinating. It was a challenge, and meeting the people – people like Barry Drayton, and all the Drayton family, and meeting Murray Tyrrell and going to his home.

Ryan continued to gain further winemaking expertise, studying with Brian Croser, a luminary of the South Australian wine industry, at new tertiary education facilities at Charles Sturt University, Wagga Wagga. And in 1977, when the winemaker at Mount Pleasant resigned, Ryan applied for, and received, the position – to the initial alarm of his wife Sylvia, then raising their young family on Sydney's northern beaches, but for whom Pokolbin became home.

From his first Mount Pleasant vintage in 1978, and through the period he described as 'white wines in their full cry', Ryan produced definitive wines – such as Elizabeth Semillon, a tremendously popular brand from the 1980s. Elizabeth Semillon embodied wines that suited the maturing of national taste for wine that occurred after the white wine boom of the '70s. As Brian McGuigan recognised, after people have been encouraged to learn about wine, many will seek out more complex styles: those that are drier, with clearer subtleties and variances of taste.[25] Ryan was pleased with Elizabeth Semillon. 'It was on every wine list you could think of.'

Ray Kidd recalled that 'consumers were now another generation on, newer people were coming in with new ideas, and greater prosperity in the country, and they wanted vintage, varietal, district wines'. And they rejected 'commercial branded non-vintage, non-district, non-varietal wines' like Ben Ean.[26]

In the dance between tradition and innovation that characterises the contemporary history of the Hunter wine community, Ryan lovingly restored the massive, almost century-old oak barrels for maturing red wines that Maurice O'Shea had purchased when he established Mount Pleasant's winery. Until his retirement in 2012, Ryan maintained a barrel of sherry begun during O'Shea's time and continuing through Brian Walsh's tenure, which Ryan liked to remind drinkers had been topped up with wine, Spanish Solera style, since the 1920s. Ryan received hundreds of international accolades for his wines (particularly Semillon) from McWilliam's vineyards at Pokolbin.

Len Evans, 'Cellarmaster' and Rothbury Estate (Pokolbin)

BORN IN ENGLAND IN 1930, LEN EVANS TRAVELLED TO NEW ZEALAND during his early twenties, lived for a while in Singapore, and worked peripatetically in Australia, before settling into food and beverage management at the Chevron Hilton Hotel in Sydney. Evans' timing was perfect. Eating and drinking in Australia were being revolutionised, not only at the everyday level of drinking dry table wines at home, but also through the revival of fine dining in the evenings, now that wine could be served after 6 pm. Hotel restaurants, such as the Hilton's, were leaders in a new era of fine dining.

Evans recollects the wine cellar at the Hilton as the best in Australia in the late 1950s. 'There's no question of that', he claimed. 'Best cellar I think that's probably ever been bred in Australia. We had huge table wines sales because we had a huge restaurant. We had the 650 seat Silver Spade, an 800 seat ballroom, an 800 seat lounge, and consequently they had a lot of wine.'[27] As a wine customer, 'I became the darling of the [wine] industry as I sold lots and lots of their wine'. Representatives of wine companies, 'Lindeman's, Penfold's, Seppelt's, Orlando, Thomas Hardy (very important to me), all very much sought my company – and Yalumba – because they had wine to sell me, and they could see that I was going in a direction in which they weren't very strong'; that is, table wine. (Most of these companies were from South Australia.)

In 1962, Evans began writing *The Bulletin*'s first wine column, as 'Cellarmaster', which he believed to be Australia's first regular discussion about wine. (As distinct from, say, the work of wine writer Walter James, who published several books from the 1950s but wrote only intermittently for the press). Soon afterwards, Evans received an invitation to join the National Promotions Executive for the Australian Wine Board, which led him to establish a program to educate people about how to select and serve wine. Tourism in regions like the Hunter became part of learning how to buy and drink wine. Publications arising from Evans' campaigns included Frank Margan's books, *The Grape and I* (1969) and *A Guide to the Hunter Valley* (1971) – the latter of which spurred self-drive tours of the region – and *The Hunter Valley: Its wines, people and history* (1973).

> Some visitors to the Hunter wanted to own a part of it, and to make their own wine. Len Evans established a syndicate of investors to seed Rothbury Estate.

Some visitors to the Hunter wanted to own a part of it, and to make their own wine. In the 1960s, as capital flowed in the direction of the rise in wine profits, Len Evans established a syndicate of investors to seed Rothbury Estate. 'It started very simply', said Evans, 'because Murray

Tyrrell said to me, "I'm going to cut up the home paddock and put some vineyards in" to sell to "Sydney people ... who want to buy a vineyard". So I said, "Well, don't do that, form a syndicate". He said, "What's a syndicate?". I said, "I don't know". With the inception of Rothbury, Evans brought his hotelier's experience to combining wine tourism and entertainment. 'I used my, if you like, entrepreneurial flair there to do things that are still talked about today, the big dinners, the ribbon dinners, the operas, the operettas. All those things that happened brought thousands of people to the Valley. People loved them', according to Evans.

The Rothbury syndicate of multiple investors represented a new corporate period for the Hunter. As distinct from wine companies that had emerged from family wine businesses into leviathans of the industry, corporations with other product bases now sought opportunities to buy into the region.

Corporations rising (again)

AUSTRALIA'S LARGE WINE COMPANIES BUILT FROM SMALL FAMILY FIRMS appeared at the turn of the 20th century, as discussed in chapter 7. Economist Kym Anderson defines the years between 1965 and 1976 as a boom in the Australian wine industry. This time, as wine profits increased, wine firms merged and the industry attracted larger investors who were not traditionally wine producers. Anderson identifies the principal mergers and acquisitions touching on the Hunter in this decade as follows. The 1969 purchase of Hungerford Hill by a consortium of lawyers and accountants, and of J Y Tulloch Wines (Glen Elgin) by British paper manufacturer Reed Consolidated. (In 1973, British spirit manufacturer Gilbey's purchased 60 per cent of J Y Tulloch Wines, and the remainder in 1976.) The 1970 purchase of Chateau Reynella by Hungerford Hill, and of EB Drayton by Pokolbin Winemakers. The 1971 purchase of Lindeman's – which had in 1965 invested in Rouge Homme, South Australia – by the US tobacco company Philip Morris, and of Hungerford Hill and Chateau Reynella by the British tobacco company Rothman's. The 1974 purchase of Elliott's Wines by Hermitage Wines and of Saxonvale at Broke by

Poster for first Hunter Vintage Festival 1973, by R. Coulson. Courtesy of R. Coulson.

CIVILISED DRINKING IN A WINE NATION

British mercantile company Ryecroft, Gollin & Co. (In 1975, the Australian brewery company Tooth's bought Penfold's Wines, which had by now divested its Hunter properties.)[28]

This investor boom period, on the one hand, provided new employment within the wine community. Jay Tulloch remained in the industry as an employee of first Reeds and then Gilbeys and later Penfolds, followed by Southcorp until 1996, after which he formed his own JYT Wine Company until such time as he joined a business partnership to reclaim the right to trade under his family name. The broader Hunter wine community grew too, through the need for managerial staff, and more cellar hands and vineyard workers.

Mechanical harvesting (not in the Hunter, location unknown). Val Foreman Photography. National Archives of Australia, B942 [WINE], 30779460_002.

Ivan Howard, Somerset Views (Pokolbin)

THE SURGE IN INTEREST FOR WINES BROUGHT LANDOWNERS IN THE Hunter back into the grape-growing fold. Some vineyard replantings began in the Hunter in the 1950s, inscribing anew the gentle contours of the broad geography of the Hunter with neat angular acreages of espaliered vines – but the pace quickened in the 1960s as existing wine families and companies in the Hunter began extending their vineyard acreages.

Ivan Howard was born in Cessnock in 1923, and grew up on his family's farm at Pokolbin, near the vineyard of the Stevens family. Howard remembers his father tallying pickers' buckets at Stevens' to determine how much the harvesters were paid.

> Every picker had a number and they'd sing out their number and the number of buckets they were carrying, usually two buckets each, and they were booked down by Dad ... We managed to get time off school to do a bit of grape picking as kids. A penny a bucket.

The buckets were 'two-and-a-half gallon' sized. 'When I was a child I only picked about 60 buckets, but the men that were running flat out picking ... would pick up to 200 buckets. Today they don't pay them by the bucket, they pay them by the hour'.[29] (Indeed, now most grapes are harvested mechanically.)

In the early 1960s, now with considerable experience of his own gained from pruning and other tasks at Stevens', Howard received a visit from the manager at McWilliam's Mount Pleasant, who suggested that Howard plant grapevines at Somerset (as he refers to the property). 'Brian Walsh came over and had a look at the soil and said yes, it would grow good grapes.

'I got cuttings from my neighbour George Stevens' father Albert while they were pruning. I went in and selected the cuttings', Howard continued, 'and carried them down to the shed and they were there ready to plant'.[30] Somerset's plantings were initially 6 acres of Shiraz and then 6 acres of Semillon. A viticulturalist from the Department of Agriculture

Tea-break for pickers at Lindeman's vineyard, *Australian Women's Weekly*, 21 April 1971.

assisted by pegging out the vineyard for free, to account for water run-off during heavy rainfall, and other factors.

While Department of Agriculture support for viticulture and winemaking had infrastructure elsewhere in the state, viticultural officer Graham Gregory is recognised as an integral figure in providing ready expertise in the Hunter in the revival of vineyard plantings.[31]

Stockhausen recalls visiting the Stevens' vineyard, next door to Howard's Somerset:

> when I first came here, they were making wine, very primitive, in a tin shed, in a cask, and we would buy it. And we realised that the grapes were of high quality. So we said, look, can we do this differently? When the grapes are ready, you pick them and bring them into us, and we'll make your wine in our cellar. And then we'll pay you for it, probably more than we can pay you now for what you've made here. And they were happy to do that, and did

that year after year. There were no signed contracts or anything. But we would make the best wine possible out of that. They would get paid a better price to start with, and then when you came to the finished wine, you would say, okay, what quality is it really? And where do we put it? In the high class wine? ... And they were very very happy with that.[32]

Because Stevens supplied Lindeman's, when the Somerset vines came into bearing, Howard decided he too would sell to this company, on the basis of a verbal agreement confirmed with only a handshake 'There were different methods of pruning, and we tried several methods. Got good crops ... Twenty-seven years I supplied Lindeman's with grapes', reminisced Howard. 'I increased the size of my vineyard year by year', working alongside his son, Glendon, who continues to manage the vineyard today.

Brian McGuigan and the Upper Hunter revival

BRIAN MCGUIGAN WAS BORN IN THE HUNTER IN 1942, WHEN HIS FATHER Perc managed Penfold's Dalwood. Interested in joining the industry, McGuigan first served an apprenticeship with Penfold's, and then went away to study at Roseworthy College with a new generation of young men from other wine families.

> I was in the era of Todd Combet [father of politician Greg Combet] and Harold Davoren from Griffith. Todd Combet, of course, from Minchinbury [in Sydney]. And a number of people at the same time. Karl Lambert as well, who was at Magill [in South Australia]. So we all went through the system together. And so I eventually ended up at Magill working for Max Schubert [Penfold's winemaker in this era] and working for Ray Beckwith as well. And then came back to the Hunter in the mid 60s as a winemaker at Dalwood, and then got involved in the development of the new Penfold property in the Upper Hunter at a place called Wybong.

Brian married his childhood sweetheart, Fay, and together they moved to Wybong in 1966. Penfold's established the site as an estate with cottages for the workers and their families; matching fibro bungalows clustered a short walk from the winery. At the same time, Perc purchased Dalwood when Penfold's offered it for sale in the 1960s. Brian and Fay continued to live at Wybong until Brian formed a consortium to purchase Dalwood from his father in the early 1970s.

During the boom in Upper Hunter vine plantings, McGuigan in turn encouraged others to the region, including John and Mary Muddle. John Muddle had great success distributing poker machines before he and Mary moved their young family to the river flats of the Hunter between Denman and Muswellbrook, to plant vines at Richmond Grove, a name, like Rosemount, synonymous with the boom in wine from the region. In a new era for women, Mary Muddle played a key role in the vineyard business, as did other women in the Hunter wine community.

In 2000, Lindsay Francis interviewed Brian McGuigan about what lay at the heart of his work in the wine industry. 'Yeah. Well, I don't know whether I can explain that', he began. 'It's just I suppose a commitment. Because I love what I do. I know my wife loves it. My daughter [Lisa] loves it. I know my father loved it. It's in your blood I suppose.'³³

> I love what I do. I know my wife loves it. My daughter loves it. I know my father loved it. It's in your blood I suppose.

'But at the same time', he ventured:

> I think it's true to say, Lindsay, that the wine industry is enchanting. And why is it? Because it's an agricultural industry in the first place. Therefore, you have the complexity of the seasonality. But then you take the product to the next phase, which is a chemical process, almost. And then it moves from that to the third phase, which is packaging it and bottling it. And then the fourth phase, and the fifth phase, of taking that product out to the consumer, understanding what it's all about, delivering, so that it appeals at the right price in the right way etc etc. So, I suppose it's the complexity

At Wyndham Estate in the early 1970s (left to right) Vanessa, Brian, Fay and Lisa McGuigan. McGuigan Family Collection.

of it and the ever-changing seasonality of it, that gets me engrossed. Because you're never doing the same thing any day. Because you're changing as the season goes on. But then I suppose I'm fortunate. And I guess my father was fortunate. We're very fortunate because we have fallen into – sort of fallen into – a career that we love. I think that I see so many people around me, who are in other businesses and activities, who are square pegs in round holes. So I know from my way of thinking, and my wife's attitude, that we just love what we do. I mean, we work at it basically seven days a week, every day of the year. Because it's not a job. It's not a career. You know, the wine business is our life. And I'm passionate about that because I love what I do.[34]

Fay McGuigan worked alongside Brian throughout their career in the wine industry, first in promoting Upper Hunter wineries with bus tours in the 1970s, and from the 1980s was instrumental in creating the export connections that propelled McGuigan's wines to significance during the second wave of wine globalisation.

Keith Yore, Verona (Denman)

KEITH YORE WAS ANOTHER KEY FIGURE IN THE RAPID GROWTH OF UPPER Hunter vineyards and wine production. He was born in Brisbane in 1921 and moved to Denman in the Upper Hunter in the early 1950s, when he and his wife Georgina bought a farm supply business. Yore's engineering degree gave him the skills to design and install irrigation equipment. Unlike the Lower Hunter vineyard district – which had no means of irrigation apart from rainfall until the 1990s, with the development of a

The oldest of these casks at a Wybong winery were German and bear the date 1831. National Archives of Australia, B941 WINE/INSPECTION/1.

private irrigation pipeline from the Hunter River – Upper Hunter winegrowers could pump directly from the river. The capacity to irrigate greatly improved the reliability of grape yields and, nationally, irrigated vineyards dominated acreages of non-irrigated vines. Yore's business installed irrigation for Robert Oatley's enterprise at Rosemount (where Chris Hancock and Philip Shaw established their wine industry careers) and at sites such as Denman Estate. Yore estimates that with his adaptation of Israeli drip-irrigation technology, his business outfitted 90 per cent of the Upper Hunter's vineyards with piping to water the vines.[35]

Yore planted his own 50 acre vineyard, Verona, to demonstrate his irrigation system. The wine from these vines went on to win a trophy in Adelaide for Rhine Riesling, in addition to many other prizes, as the Upper Hunter established a reputation for red and white wines – excelling with white wines in particular.

After the global financial crisis in 2008, the Upper Hunter experienced the most dramatic contraction in vine acreage of any region in Australia, losing one-third of plantings in five years, and experiencing the lowest rate of replanting of any Australian wine subregion. The devastation of the subregion was due to a fall in Australia's export trade when the Australian dollar reached equal value with the American dollar, driving down American wine imports. Large-scale coal mining in the Upper Hunter has also brought changes to the landscape, and to small towns in the district.

Travelling the roads of the Upper Hunter with Brian McGuigan today is to relive the heyday of Yore's Verona alongside Penfold's Wybong and Richmond Grove – to reimagine the small valleys that appear at the end of long roads as broadly carpeted with thriving vineyards. Other former sites that McGuigan points out are Rosemount, Hollydene and Sobels. Although most of the Renaissance vineyards are long gone, John and Mary Muddle's daughter Linda Keeping, and her husband Brett, have nurtured Two Rivers Wines near Denman since 1992. And, behind the scenes in the repurposed Oak Dairy Factory at Muswellbrook, second generation wine businessman John Hordern (son of Bobby Hordern) is driving the export of Hunter wines to unexpected places, like Russia, as part of the tremendous diversification of wine flows that have emerged over the past half-century.

A new vineyard association for a new era

EARLIER ORGANISATIONS FOR VINEGROWERS AND WINEMAKERS IN THE Hunter had faded away mid-century, along with the winegrowing community. Then, by 1972, for the reasons discussed – and because of the absence of capital gains tax on new vine plantings – Hunter vineyard acreages had doubled, mainly around Broke-Fordwich and in the Upper Hunter. To focus the interests and advocacy of the newly expanded Hunter winegrowing community, Alec Forsythe – an accountant and managing director of Mindaribba Vineyards – called a meeting at the Pannaroo Motel, Singleton, to form a new regional group. Those in attendance included members of the Drayton family, Murray Tyrrell, the Tullochs, Brian McGuigan, Max Lake, Doug Elliott, Christopher Barnes (Saxonvale Wines), Don McWilliam (McWilliam's Mount Pleasant), Simon Currant (Hungerford Hill), David Webster (Hermitage Wine Company), Ivan Howard, Murray Robson (Squire vineyard), and Lance Allen (Tamberlaine Wines).[36]

Forsythe was elected as the first president of the Hunter Valley Vineyard Association, and was succeeded the following year by Brian McGuigan, with Christopher Barnes as association secretary. After the initial Singleton gathering the group's meeting were held in Pokolbin, and Barnes recalls that the HVVA represented almost every vineyard and wine business in the region – with as many as 100 people attending the annual general meetings. A committee of nine was elected annually, and the president formed subcommittees. One subcommittee was set up to start a Hunter Wine Show. There was a technical subcommittee, a viticultural subcommittee, a promotions committee and so on. Barnes points out too that the association worked closely with the Department of Agriculture, and organised a seminar and field day each year. In the 1990s, Barnes and other were continuing members, and the executive included Bruce Tyrrell,

Checking wine in cellars — Hunter River — New South Wales, Australian News & Information Bureau. National Archives Australia B942, WINE [4].

Richard Hilder, David Lowe, Patrick Auld and Ian Tinkler. In 2010 the HVVA merged with Wine Hunter Marketing to become the Hunter Valley Wine Industry Association. In 2016, the HVWIA amalgamated with Hunter Valley Wine Country Tourism to form the Hunter Valley Wine and Tourism Association.

Iain Riggs, Brokenwood (Pokolbin)

WHEN IAIN RIGGS ARRIVED AT POKOLBIN IN 1982 AS A WINEMAKER FOR Brokenwood, he represented a new generation for the Hunter wine community at the end of the Renaissance generation; the wine men (there would be more women in the next generation) who could not recall the lean years of temperance – men who trained in mechanised facilities, with world-renowned experts, and who would oversee the transition into the second wave of wine globalisation, of which the Australian wine industry was an integral part, from the mid-1990s.

Riggs was born in Burra, South Australia, in 1955. He describes his home town as 'about an hour and a half north of Adelaide, and if you sort of head to Burra and went east you'd end up at Broken Hill and the back blocks [of New South Wales] and if you went west you go up to Port Augusta and further. It's one of the last major towns. It's a Cornish mining town'.[37] During Riggs' teenage years his family relocated to a farm in New South Wales before returning to South Australia.

Iain Riggs, Brokenwood, Pokolbin c1983. Iain Riggs Collection.

Riggs – 'a farm kid all his life' – became steeped in South Australia's Riverland wine industry as it flowered. He left school in 1972 to study at Roseworthy College, graduating in 1975 with Second Class Honours in Oenology, all the while gaining experience at first Waikerie Cooperative winery and then Bleasdale Vineyards.

During his training at Roseworthy, Riggs' lecturers described the Hunter as 'the last place on earth you'd grow grapes', and conditions in the late 1960s and early 1970s seemed to confirm this. In 1968 there were bushfires in the Pokolbin district which destroyed many winery buildings, most notably Daisy Hill, and then the early vintages of the 1970s were washed out with heavy rainfall at grape harvest. Yet, after working for ten years at Bleasdale, Langhorne Creek, and then briefly at Hazelmere Estate, in McLaren Vale (and by now the recipient of the region's prestigious Bushing King Award), Riggs accepted a job with Brokenwood at Pokolbin. He and his wife moved east in time for him to take charge of the completion of new winery facilities for the 1983 vintage.

Brokenwood was founded in 1970 as a partnership between Sydney solicitors Tony Albert, John Beeston and James Halliday (before his stellar career as a wine writer). The first two Brokenwood vintages, in 1973 and 1974, were made into wine at the facilities at nearby Rothbury Estate. Those first grapes were 'all trucked over, transported in the back of Len Evans' Bentley in buckets', to be crushed and fermented. A two-storey building was constructed for the 1975 vintage: a downstairs winery area and tractor shed, and an upstairs bunkhouse and kitchen that was later converted into the Brokenwood cellar door. The extensions to the winery plant that Riggs oversaw in 1983 provided new state-of-the-art equipment for fermentation and storage still in use today. 'I arrived the 3rd or 4th of November 1982 and by New Year's Eve '82, '83 we had the shed up and operating in time for harvest later that month', recalled Riggs.

How did he view this welcome to the Hunter?

'It was good. Yeah, it was very good ... The '83 harvest was hot and dry but our processing really changed the style of the Semillon forever ... It was the first time that a very fragrant, very grassy Semillon had been seen out of the Hunter Valley.'

Riggs' arrived in the Hunter as a new generation of Australians who had learned to drink wine with styles such as Lindeman's Ben Ean Moselle and TR2 began to seek new sensory experiences from wine, and to differentiate their taste in wine through drier styles, wines that required greater knowledge of wine places and vintage details.

Towards the second wave of wine globalisation

THIRD GENERATION HUNTER WINEGROWER MURRAY TYRRELL IN 1983 declared that year to be a turning point for the Hunter wine region. According to Tyrrell, a primary reason for this change was the success of campaigns to reduce air pollution from the Kurri Kurri Aluminium Smelter. Vignerons had noticed discolouration and poor vine performance and analysis showed a build-up of fluoride in the grapes. New regulations on air quality cleansed the maritime air currents flowing from the sea into Pokolbin and surrounds, currents that brought tumultuous weather patterns but also the milder conditions that provided the basis of the Hunter's light wine styles, compared with other low-altitude Australian wine regions. After a 'big red' era caused by drought conditions in the early 1980s – bringing fuller body to the region's wines – the breaking of the drought after the 1983 vintage spelled a return to the lighter styles of red wine for which the Hunter was best known.[38]

Tyrrell's description of the Hunter vintage was keenly observed among a substantial number of Australian wine lovers. The new era of introduction wines had given way to a maturing culture of interest in the changing taste of wines from now-famous vineyards. A new tradition of civilised (and gender-inclusive) drinking gave the Australian wine industry a base of popular wines from which to encourage new connoisseurs of fine wines that would become a fascination for many Australians – along with gastronomic delights that were bringing excitement to eating and drinking culture. From the springboard of a new wine nationalism in the 1980s, members of the Hunter winegrowing – and wine business – community were at the vanguard of a second wave of wine globalisation, led by Australian wine producers.

> From the springboard of a new wine nationalism in the 1980s, members of the Hunter winegrowing community were at the vanguard of a second wave of wine globalisation.

As of the mid-1990s, Australia wine producers would rapidly raise the quantity of wine exported to an astonishing two-thirds of total national production. While wines from grapes in the Hunter were a mere drop in this vast ocean that newly linked Australia with the northern hemisphere, many Hunter wine people, places and stories became, and remain, symbols of Australian wine in the contemporary global marketplace.

Operating the press. Murray Tyrrell, left, Murray Flannigan, right, and a third man at centre. Tyrrell's Wines, year unknown. Photograph from Norm Barney's collection, courtesy of Mrs Daphne Barney. Source: Greg and Sylvia Ray.

Conclusion: the blood of the grape

It is late one January night in the early 1980s, and a winemaker from the Lower Hunter is driving a small farm truck with a 'dog' (a large high-sided box trailer) hitched on, hurtling eastward on the labyrinthine backroads of the Upper Hunter near Jerry's Plains. The truck's headlights and tail-lights are the only illumination in the vast eerie inkiness between earth and stars. The winemaker is watching for kangaroos, wombats, foxes or rabbits that might cross in front of him on their own late-night odyssey. He knows that if an animal jumps onto the road from either left or right, he will have only a few seconds to react and swerve to avoid a collision that might slow him down.

In the febrile excitement of industry growth, many split-second decisions are made.

The winemaker in the truck speeding across Jerry's Plains is driving by dark with the largest load of grapes that his vehicle can carry. He is avoiding the thrumming heat of the day, to ensure that any juice created by the weight of the fruit self-crushing does not begin to oxidise or ferment. The knowledge that the laws of gravity mean some berries are splitting, and the juice draining away, is pressing heavily on his mind. Despite the exactness of fermentation that is possible in enormous refrigerated tanks – from which clear and pure wines will be pumped after centrifugal processes remove the lees – the winemaker prefers the grapes to arrive in good condition at the winery.

Some 150 years earlier, Edward Parry surely felt a similar sense of anticipation as he oversaw the stomping of grapes by convicts at Tahlee – although the fruit was crushed only a short distance from where it was grown. Ludwig Leichhardt, in 1843, experienced the satisfaction of observing the mysteries of fermentation unfolding over several days after a chaotic vintage at Glendon. A century earlier than our driver made his trek, The Vagabond sat in the office of John Wyndham and admired the accolades awarded to Dalwood wines. (Some 20 years after that, Will Brecht of Rosemount spoke dismissively of the outmoded, 'Scriptural' winery equipment that he had recently replaced. What might he have made of the shiny possibilities offered by the electrified and automated processes in the winery where our driver would deliver his grapes?)

At Pokolbin in the 1940s, the two mile journey from Trevena's vineyard to Elliott's Oakdale, in a horse-drawn cart loaded with barrels of freshly picked grapes, seemed to take an eternity. Now, as our driver watches his headlights bounce with the body of the truck on the corrugations of the road, millions of bunches of grapes in refrigerated trucks and freight containers are criss-crossing the nation by road and rail.

Our driver is also taking his journey by starlight to dodge day-working 'scalies', the transport inspectors who enforce safety regulations on the weight of goods in road vehicles. The truck and trailer we are travelling with through a eucalypt-fumed night is overloaded. The winemaker is reminded of this each time he swiftly changes down gears to make the sudden turns of the road, the dog sliding from side to side on the uneven, gravelled surface. The driver could do with an extra hand to shunt the levers on the dashboard that control the trailer, but his perceptions are sharpened by the need for combined haste and precision. Truck and trailer are loaded with grapes picked on the western side of the Great Dividing Range, and the winemaker is bringing them to the Lower Hunter, to be made into wine with grapes from the district. Wine that within a year (or more) will be sitting, its contents deceptively unrippled – given the hurry and hazards of production – in a bottle, on a shelf or wine rack, almost anywhere in the country, or the world.

Stories about wine, like wine itself, are an ancient source of human pleasure. But wine is only an object – an enjoyable one, no doubt. Much is (rightly) made of the nature of the physical environment where grapes are grown. And more could be made of this environment as a whole ecology,

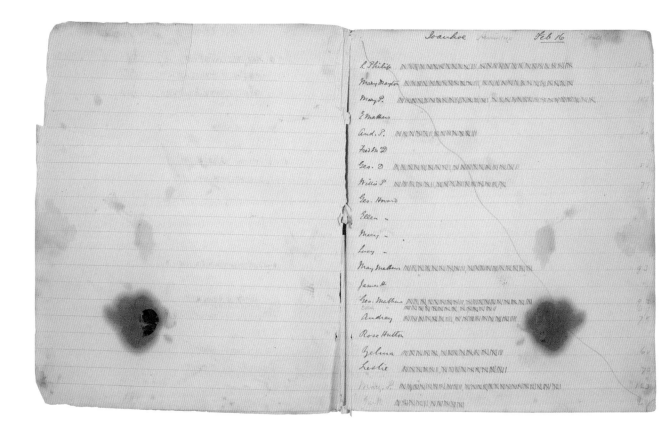

Exercise book with statistics for 1891–1892 vintages, Wilkinson Family Papers, 1836–1961, State Library of NSW, Call No. Mitchell Library 532, Box 1 (7).

and of ensuring it remains suitable for winegrowing. Yet people are the lifeblood of wine, and their stories form the threads of the fabric of the winegrowing community. Across the arc of generations of Hunter winegrowers since 1828, people have bumped up against each other, solved problems, made decisions – taken risks. In tracing the manifold circuitries of Hunter connections from a local through to a global level, we have named only some of the people who featured across six generations to 1983. Within this history many people are still nameless: the truck driver on the backroads, and grape pickers toting buckets (or in the 1980s, driving new mechanical harvesters).

In the State Library of New South Wales manuscripts collection is a tally book for the grape harvest at Lionel Viney Hindmarsh Stephens' Ivanhoe vineyard at Pokolbin, on 16 February 1891. The names on the right-hand side of the page are the men, women and (quite likely) children who laboured to gather Hermitage (Shiraz) grapes. Each of the small notches lined up beside the names represents a bucket of grapes that the picker had cut from the vine, then lugged to a dray loaded with open-topped hogshead barrels. The picker called out their name to the tally keeper holding the notebook as the full bucket was handed up to another worker, who poured the grapes into a barrel. And then the picker toted the empty bucket back to the vines, ready to be filled again.

Look closely, and preserved between the pages of the tally book you will see a small red berry, caught, crushed and colouring this record of a working day 127 years ago.

Notes

Introduction: making wine history

1. TC Harington, Colonial Secretary's Office to Charles Fraser, Botanic Gardens, 10 July 1828, letter received and recorded in Particulars of Letters received & answered in the Botanical Department of Sydney commencing 19 May 1828, The Daniel Solander Library, Royal Botanic Gardens Sydney (DSL), A1, n.p.
2. See for example, JHM Abbott, *The Newcastle Packets and the Hunter Valley*, Currawong Publishing, Sydney, 1943, ch. 16, 'Vin du Pays'.
3. Busby and Kelman Papers 1822–1879, Mitchell Library (ML) MSS 1183, State Library of NSW (SLNSW). On land title matters see Jack Sullivan, *Dr Henry Lindeman and Cawarra, Gresford*, Part I, Paterson Valley Historical Society, Paterson, 2014, p. 31. For other evidence on Busby's contribution see Julie McIntyre, *First Vintage: Wine in Colonial New South Wales*, UNSW Press, Sydney, 2012.
4. James Busby to William Kelman, 13 September 1824, Busby and Kelman Papers, ML MSS 1183, SLNSW, p. 79.
5. Nick Brodie, *1787*, Hardie Grant Books, Richmond Vic., 2016.
6. This history is explored in detail in McIntyre, *First Vintage*. See also Julie McIntyre, 'Trans-"imperial eyes" in British Atlantic Voyaging to Australia, 1787–1791', *History Australia*, 2018 (accepted for publication 23 December 2017).
7. Various letters, Busby and Kelman Papers, ML MSS 1183, SLNSW.
8. James Busby, *A Manual of Plain Directions for Planting and Cultivating Vineyards, and for Making Wine in New South Wales*, Printed by R Mansfield, for the Executors of R Howe, Sydney, 1830.
9. Diary of George Wyndham of Dalwood, 1830 – February 1840, ML MSS B1313, SLNSW, n.p.
10. James Busby, *Journal of a Tour through Some of the Vineyards of Spain and France* (1833), Facsimile edition, David Ell Press, Sydney, 1979.
11. The description of wine styles and classes is taken from John Gunn and RM Gollan, *Report on the Wine Industry of Australia*, Government Printer, Canberra, 1931, p. 1.
12. *Singleton Argus*, 7 November 1952.
13. Gunn and Gollan, *Report*, p. 10.
14. Gunn and Gollan, *Report*, p. 10.
15. See RD Townsend (ed.), *Rambles in New South Wales*, Sydney Grammar School Press, Sydney, 1996; and Ian Grantham (ed.), *XYZ Goes North*, Convict Trail Project Occasional Monographs, Wirrimbirra Workshop, Kulnura NSW, 1999.
16. XYZ, 'An Account of a Trip to Hunter's River', *The Australian*, 7 February 1827. On young grape vines (likely for eating, not wine) 'trampled' on the property of Matthew Chapman at Williams River in 1826, see *The Monitor*, 23 June 1826.
17. XYZ, 'An Account of a Trip to Hunter's River', *The Australian*, 31 January 1827.
18. XYZ, 'An Account of a Trip to Hunter's River', *The Australian*, 3 February 1827.
19. XYZ, 'An Account of a Trip to Hunter's River', *The Australian*, 3 February 1827.
20. XYZ, 'An Account of a Trip to Hunter's River', *The Australian*, 10 February 1827.
21. Journal kept by James Macarthur at Port Stephens in pencil and written in ink about a week afterwards at Sydney, received in London 17 November 1828, Australian Agricultural Company Records (AA Co) [78/1/6], Noel Butlin Archives Centre (NBAC).
22. John Macarthur to the Governor and Directors, Port Stephens, Despatch No. 1, 26 May 1828 [78/1/6], NBAC.
23. List of Plants Cultivated in the Botanic Gardens, Sydney, New South Wales, January 1828, DSL B1, p. 512.
24. Phillip O'Neill, 'The Gastronomic Landscape', in Phil McManus, Phillip O'Neill and Robert Loughran and OR Lescure, (eds), *Journeys: The Making of the Hunter Region*, Allen & Unwin, Sydney, 2000, pp. 158, 178.
25. Robert E White, *Understanding Vineyard Soils*, Oxford University Press, Oxford, 2015, p. 225.
26. Marion Demossier, *Burgundy: A Global Anthropology of Place and Taste*, Berghahn, Oxford, 2018.
27. Julie McIntyre, 'Wine Worlds are Animal Worlds too: Native Australian Animal Vine Feeders and Interspecies Relations in the Ecologies that Host Vineyards', in Nancy Cushing and Jodi Frawley (eds), *Animals Count: How Population Size Matters in Animal–Human Relations*, Routledge, London, 2018.
28. Brian J Sommers, *The Geography of Wine: How Landscapes, Cultures, Terroir, and the Weather Make a Good Drop*, Plume, New York, 2008.
29. Daniel Honan, '"The Elements Are Against Us": Hunter Valley wine and the paradox of an enduring relationship between people and environment', Honours thesis, History, University of Newcastle, 2017.
30. Julie McIntyre, Rebecca Mitchell, Brendan Boyle and Shaun Ryan, 'We Used to Get and Give a Lot of Help: Networking and Knowledge Flow in the Hunter Valley Wine Cluster', *Australian Economic History Review*, vol. 53, no. 3, 2013, pp. 1–21.

31 An expanded version of this review of literature is provided in Julie McIntyre, 'A "Civilized" Drink and a "Civilizing" Industry: wine growing and cultural imagining in colonial New South Wales', PhD thesis, History, University of Sydney, 2009.

32 W P Driscoll, *The Beginnings of the Wine Industry in the Hunter Valley*, Newcastle History Monographs No. 5, Newcastle Public Library, The Council of the City of Newcastle, Newcastle, 1969.

33 See Julie McIntyre and John Germov, 'The Changing Global Taste for Wine: An Historical Sociological Perspective', in John Germov and Lauren Williams (eds), *A Sociology of Food and Nutrition: The Social Appetite*, Oxford University Press, Melbourne, 2017, pp. 202–18.

34 Julie McIntyre, *First Vintage: Wine in Colonial New South Wales*, UNSW Press, Sydney, NSW, 2012, see esp. chs 10 and 11; McIntyre, Mitchell, Boyle and Ryan, 'We Used to Get and Give'.

35 Julie McIntyre and John Germov, '"Who Wants to be a Millionaire? I do": Postwar Australian Wine, Gendered Culture and Class', *Journal of Australian Studies*, vol. 42, no. 1, 2018, pp. 65–84.

36 Wine giant, Treasury Wine Estates, faces takeover battle', *Bush Telegraph*, ABC Radio National, 3 June 2014, <http://www.abc.net.au/radionational/programs/bushtelegraph/twe-takeover/5497378> (accessed 6 January 2018).

37 This method may be summed up as a connected series of asymmetrical micro-inquiries that answer questions about the histories of cultural forms, such as music where the facets of folklore, performance and so on need to be understood discretely and then re-entangled in a larger study. See Michael Werner and Bénédicte Zimmermann, 'Beyond Comparison: Histoire croisée and the challenge of reflexivity', *History and Theory*, vol. 45, February 2006, pp. 30–50.

38 Legally, wine regions are defined by 'geographical indications'; see Michael Blakeney, *The Protection of Geographical Indications: Law and Practice*, Edward Elgar, Cheltenham, 2014. Historically and sociologically a 'wine region' is self-conscious and continually assessed and represented in an ongoing conversation between its interlinked production, trade and consumption communities across generations. We call this the 'wine complex' to distinguish the web of connections from the linear value chain described by economists. Our historical sociological methodology is informed by the dynamic interactions of agents within (changing) structures, as first described in Julie McIntyre and John Germov, 'Drinking History: Enjoying Wine in Early Colonial New South Wales', in S Eriksson, M Hastie and R Roberts (eds), *Eat History: Food and Drink in Australia and Beyond*, Cambridge Scholars Press, Melbourne, 2013, pp. 120–42. We further developed this concept in McIntyre and Germov, 'Who Wants to be a Millionaire?'

Chapter 1 *Vitis vinifera* in Aboriginal Country: introducing an ancient plant to an ancient land

1 Patrick E McGovern, *Ancient Wine: The Search for the Origins of Viniculture*, Princeton University Press, Princeton, 2003.

2 When Indigenous people began to live in Australia remains unresolved. This is the most recent research on this question: Chris Clarkson, Zenobia Jacobs … Colin Pardoe, 'Human Occupation of Northern Australia by 65,000 years ago', *Nature*, vol. 547, 2017, pp. 306–10. See also R Tobler et al, 'Aboriginal Mitogenomes reveal 50,000 years of Regionalism in Australia', *Nature*, vol. 544, 2017, pp. 180–84.

3 On the notion of nested geographical spaces within the Hunter Valley see for example Mark Dunn, 'A Valley in a Valley: Colonial struggles over land and resources in the Hunter Valley NSW 1820–1850', PhD thesis, History, University of NSW, 2015.

4 On Hunter geography see P Shimeld, R Drysdale, R Loughran, 'The Physical Landscape of the Hunter Valley', in *Journeys*, p. 12. See historicisation of Patrick's Plains within the wider Hunter in Dunn, 'A Valley in a Valley'.

5 Shimeld, Drysdale, Loughran, 'The Physical Landscape of the Hunter Valley,' p. 25.

6 Bill Hicks, 'Wollombi Landcare', in Denis Mahony and Joe Whitehead (eds), *The Way of the River, Environmental Perspectives on the Wollombi*, Wollombi Valley Landcare Group and The University of Newcastle, Newcastle, 1994, p. 25.

7 See Dunn, 'A Valley within a Valley'. Also Helen Brayshaw, *Aborigines of the Hunter Valley: A Study of Colonial Records*, Scone & Upper Hunter Historical Society, Scone, 1986; James Miller, *Koori, A Will to Win: the heroic resistance, survival & triumph of Black Australia*, Angus & Robertson, Sydney, 1985.

8 Bill Gammage, *The Biggest Estate on Earth: How Aborigines Made Australia*, Allen & Unwin, Sydney, 2011, p. 61.

9 Gammage, *Biggest Estate*, pp. 91–93.

10 Grace Karskens, 'Fire in the forest? Exploring the human–ecological history of Australia's first frontier', *Environment and History*, 2019, forthcoming.

11 Peter Cunningham, *Two Years in New South Wales: A series of letters comprising sketches of the actual state of society in that colony, of its peculiar advantages to emigrants, of its topography, natural history etc*, vol. 1, Printed by Henry Colburn, London, 1827, p. 156.

12 XYZ, 'An Account of a Trip to Hunter's River', *The Australian*, 7 February 1827.

13 Cunningham, *Two Years in New South Wales*, pp. 151–53. Cunningham's emphasis.

14 James Atkinson, *An Account of Agriculture and Grazing in New South Wales and some of its most useful natural productions, with other information important to those who are about to emigrate to that country, as a result of several years residence and practical experience*, Printed for James Atkinson, London, 1826, pp. 13–17.

15 Mark Dunn, 'Aboriginal guides in the Hunter Valley, New South Wales', in Tiffany Shellam, Maria Nugent, Shino Konishi and Allison Cadzow (eds), *Brokers and Boundaries: Colonial Exploration in Indigenous Territory*, ANU Press, Canberra, 2016, pp. 61–84.

16 Tamsin O'Connor, 'Charting New Waters with Old Patterns: Smugglers and Pirates at the Penal Station and Port of Newcastle 1804–1823', *Journal of Australian Colonial History*, vol. 19, 2017, p. 1.

17 David A Roberts and Daniel Garland, 'The Forgotten Commandant: James Wallis and the Newcastle Penal Settlement, 1816–1818', *Australian Historical Studies*, vol. 41, no. 1, 2010, p. 9.

18 Brian Walsh, *European Settlement at Paterson River 1812 to 1822*, Paterson Historical Society, Paterson, 2012, pp. vi, 5–6.

19 JF Campbell, 'The Genesis of Rural Settlement on the Hunter', *Journal of the Royal Australian Historical Society*, vol. 12, no. 2, 1926, p. 74.

20 XYZ, 'An Account of a Trip to Hunter's River', *The Australian*, 17 February 1827.

21 Charles Ludington, *The Politics of Wine in Britain: A New Cultural History*, Palgrave Macmillan, London, 2013.

22 James Atkinson, *An Account*, p. 7. A posthumous reprinting of *An Account* (1844) contains a short boosterist chapter on viticulture borrowed from other sources and signals the importance of this branch of agriculture at this time.

23 James Atkinson, *An Account*, p. 8.

24 On Atkinson and the colonial Hawkesbury see Grace Karskens, 'The Settler Evolution: Space, Place and Memory in Early Colonial Australia', *Journal of the Association for the Study of Australian Literature*, vol. 13, no. 2, 2013, pp. 1–21. Also, Grace Karskens, *People of the River*, Allen & Unwin, Sydney, forthcoming, ch. 9.

25 Peter Cunningham, *Two Years in New South Wales*, p. 147.

26 *Maitland Mercury*, 22 August 1827, p. 3.

27 Atkinson, *An Account* (statement), p. 34 and (quotations), pp. 56–57.

28 Atkinson, *An Account*, 59.

29 McIntyre, *First Vintage*.

30 On climate parity see David Dunstan and Julie McIntyre, 'Wine, olives, silk and fruits: The Mediterranean plant complex and agrarian visions for a "practical economic future" in colonial Australia', *Journal of Australian Colonial History*, vol. 16, 2013, pp. 29–50.

31 Pennie Pemberton, 'The London Connection: the formation and history of the Australian Agricultural Company 1824–1834', PhD thesis, History, Australian National University, 1991, p. 163.

32 John Jamison, *Report to Agricultural Society of New South Wales*, 23 February 1826, Sydney, p. 2.

33 John Connor, *The Australian Frontier Wars 1788–1838*, UNSW Press, Sydney, 2002. Thanks are due to Lyndall Ryan for guidance on this matter. See also Kelly K Chaves, '"A Solemn Judicial Farce, the Mere Mockery of a Trial": The Acquittal of Lieutenant Lowe, 1827', *Aboriginal History*, vol. 31, 2007, pp. 122–40.

34 XYZ, 'An Account of a Trip to Hunter's River', *The Australian*, 7 February 1827.

35 We are grateful to Jill Barnes and Peter Read for guidance on this matter, and for advancing our understanding on the most respectful way to discuss the borders between Aboriginal tribal groups.

36 See Julie McIntyre, '"Bannelong Sat Down to Dinner with Governor Phillip, and Drank his Wine and Coffee as Usual": Aborigines and Wine in Early New South Wales', *History Australia*, vol. 5, no. 2, 2008, pp. 39.1–39.14.

Chapter 2 A network of instigators: imagining Hunter winegrowing, 1828–35

1 Cameron Archer surmises the Line Paddock vineyard trenching occurred in the 1820s and 1830s. A Cameron Archer, 'Social and Environmental Change as Determinants of Ecosystem Health: A Case Study of social ecological systems in the Paterson Valley, NSW, Australia', PhD thesis, Life and Environmental Sciences, University of Newcastle, 2007, pp. 117, 190.

2 The State Library of New South Wales (SLNSW) holds eight copies of James Busby, *Manual*. Hely's copy of the *Manual* is in the Dixson Library (DL) collection, call no. 634.8/B copy 1.

3 Driscoll, *Beginnings*, pp. 11–12.

4 The list of members of the Agricultural Club is from *The Australian*, 22 August 1827.

5 Brian Walsh, 'Heartbreak and Hope, Deference and Defiance on the Yimmang: Tocal's convicts 1822–1840', PhD thesis, History, University of Newcastle, 2007, p. 11.

6 Walsh, 'Tocal's convicts 1822–1840', p. 103.

7 Archer, 'Social and Environmental Change', pp. 62, 117.

8 Driscoll, *Beginnings*, p. 28.

9 Brian Walsh, *Voices from Tocal: Convict Life on a Rural Estate*, CB Alexander Foundation, Paterson, 2008, p.114; Driscoll, *Beginnings*, p.19.

10 <http://adb.anu.edu.au/biography/scott-robert-2642> (accessed 7 December 2017).

11 <http://adb.anu.edu.au/biography/cory-edward-gostwyck-1922> (accessed 8 January 2018).

12 Walsh, *Voices*, p. 77.

13 Lesley C Gent, *Gostwyck, Paterson 1823–2009*, Paterson Historical Society, Paterson, 2009, p. 13; Miller, *Koori, Will to Win*.

14 <http://www.patersonriver.com.au/people/langganda.htm> (accessed 7 December 2017).

15 JD Lang, *A Historical and Statistical Account of New South Wales*, Fourth edition, Printed by Sampson Low, Marston, Low & Searle, London, 1875, p. 106.

16 Lang, *A Historical and Statistical Account* (1875), p. 244.

17 Julie McIntyre, 'Adam Smith and Faith in the Transformative Qualities of Wine in Colonial New South Wales', *Australian Historical Studies*, vol. 42, no. 2, 2011, pp. 194–211.

18 Cunningham, *Two Years in New South Wales*, p. 157. Emphasis in original.

19 <http://adb.anu.edu.au/biography/cunningham-peter-miller-1942/text2325> (accessed 1 October 2017).

20 Letter from Ellen Ogilvie cited in W Allan Wood, *Dawn in the Valley: The Story of Settlement in the Hunter River Valley to 1833*, Wentworth Books, Sydney, 1972, p. 188.

21 There are no details on winegrowing in the Little Family Papers 1838–1863, ML MSS 1671, SLNSW.

22 <http://adb.anu.edu.au/biography/dumaresq-william-john-2239> (accessed 18 March 2018).

23 Henry Dumaresq to his mother, 26 March 1827, ML MSS A2571, SLNSW.

24 Henry Dumaresq to Charles Winn, 26 May 1831, ML MSS A2571, SLNSW.

25 Introductory notes, Dumaresq Papers, ML MSS A2571, SLNSW.

26 Jack Sullivan, *George Townshend 1798–1872 and Trevallyn, Paterson River*, Paterson River Historical Society, Paterson, 1997, pp. 7–10, 21.

27 Wyndham, Diary, 1830–1840, ML MSS B1313, SLNSW, n.p.

28 Charles Boydell, Journal, 1830–1835, ML A 2014, SLNSW, entry for January 1833.

29 Wood, *Dawn in the Valley*, pp. 65, 162–64. Note that Wright received vine cuttings from the Botanic Gardens, Sydney, in 1829, as did Alexander McLeod at Luskintyre and James King, B1, DSL, pp. 59–60. These appear to be small quantities of vines. Our thanks to Don Seton Wilkinson for these details.

30 Wyndham, Diary, 1830–1840, ML MSS B1313, SLNSW.

31 *Sydney Gazette*, 20 May 1830.

32 *Maitland Mercury*, 3 June 1843.

33 Cynthia Hunter (ed.), *The 1827 Newcastle Notebook and Letters of Lieutenant William S. Coke, H.M. 39th Regiment: Officer in charge of the military garrison stationed at Newcastle during 1827*, Hunter House, Raymond Terrace, 1997, p. 52. See also Helenus Scott to William Kelman, 28 November 1833, Busby and Kelman Papers, ML MSS 1183, SLNSW, p. 271.

34 Paul Struan Robertson, *Proclaiming 'Unsearchable Riches': Newcastle and the Minority Evangelical Anglicans, 1788–1900*, Centre for the Study of Australian Christianity, Sydney, 1996, p. 32.

35 James Busby to William Kelman, 13 September 1824, Busby and Kelman Papers, ML MSS 1183, SLNSW.

36 XYZ, 'An Account of a Trip to Hunter's River', *The Australian*, 7 February 1827.

37 Busby, *Manual*, pp. 41–43.

38 Julie McIntyre, 'Camden to Paris and London: The Role of the Macarthur Family in the Early NSW Wine Industry', *History Compass*, vol. 5, no. 2, 2007, pp. 427–38.

39 Lang, *An Historical and Statistical Account* (1875), p. 244.

40 Miller, *Koori: A Will to Win*.

41 Wyndham, Diary, 1830–1840, ML MSS B1313, SLNSW; McIntyre, *First Vintage*, pp. 197–215.

42 <http://www.historyofaboriginalsydney.edu.au/north-west/working-land-%E2%80%93-more-early-family-history-gavi-duncan> (accessed 28 February 2018).

43 McIntyre, Mitchell, Boyle and Ryan, 'We Used to Get and Give'.

44 <http://adb.anu.edu.au/biography/busby-john-1859> (accessed 17 October 2017).

45 William Kelman to his Mother, 8 December 1823 [on board the *Triton*] Busby and Kelman Papers, ML MSS 1183, SLNSW, p. 62.

46 Notes in Busby and Kelman Papers, ML MSS 1183, SLNSW.

47 DN Jeans, *An Historical Geography of New South Wales to 1901*, Reed Education, Sydney, 1972, pp. 32–33, 186.

48 Catherine Kelman to George Busby, 13 July 1829, Busby and Kelman Papers, ML MSS 1183 SLNSW, p. 211.

49 See Alexander Busby to William Kelman, 10 June 1833, Busby and Kelman Papers, ML MSS 1183, SLNSW, p. 234.

50 William Busby to Catherine Kelman, 28 July 1833, Busby and Kelman Papers, ML MSS 1183, SLNSW, p. 241.

51 Busby to Kelman, 28 July 1833, Busby and Kelman Papers, ML MSS 1183, SLNSW.

52 R Wood [of St Peters] to William Kelman, 31 July 1833, Busby and Kelman Papers, ML MSS 1183, SLNSW, pp. 243–44.

53 William Busby to William Kelman, 18 November 1833, Busby and Kelman Papers, ML MSS 1183, SLNSW, pp. 267–69.

54 Walsh, *Voices*.

55 Blackman [Only author detail], Catalogue of the remaining part of the library of James P Webber Esq, Pamphlet printed by Stephens and Stokes, Sydney, 1835, ML018.2W, SLNSW. Provided by Cameron Archer.

56 M Duhamel du Monceau, *Elements of Agriculture*, vol. 1, transl. by Philip Miller, Printed for P Vaillant and T Durham, London, 1764, pp. 102–103.

57 Biographical details from Walsh, *Voices*, pp. 20–21.

58 See David Hancock, *Citizens of the World: London Merchants and the Integration of the British Atlantic Community, 1735–1785*, University of Michigan Press, Ann Arbor, 1998.

59 Lang, *An Historical and Statistical Account, Volume 2* (1875), p. 33. As with other references in Lang's subsequent editions, the ideas that emerged later are inserted into the book's revised editions.

Chapter 3 Experiments and expectations: willing Hunter wine into being, 1836–46

1. James King to Lord Grey, Secretary of State for the Colonies, 26 September 1849, in AA Company papers [78/1/19, p. 425], NBAC.
2. J Carmichael, 'Irrawang vineyard and pottery, east Australia' [picture], c1838, Rex Nan Kivell Collection NK1477, National Library of Australia (NLA).
3. <https://www.jenwilletts.com/william_caswell.htm> (accessed 1 December 2017).
4. Edward Parry to Court of Directors in London, Despatch 7, 19 October 1829 [78/1/9, p. 25], NBAC.
5. Sir Edward Parry's Diary, 3 February 1832, ML A630, SLNSW.
6. Busby, *Manual*, p. 90.
7. Parry's Diary, 3 February 1832, ML A630, SLNSW.
8. Parry's Diary, 4 February 1833, ML A630, SLNSW.
9. Quote from Edward Parry to AA Company Court of Directors, Despatch 119, 8 January 1834 [78/1/13], NBAC.
10. Parry's Diary, 15 April 1834, ML A630, SLNSW.
11. Lang, *A Historical and Statistical Account, Volume 2* (1834), pp. 351–52.
12. <http://adb.anu.edu.au/biography/carmichael-henry-1881> (accessed 7 December 2017).
13. Julie McIntyre, 'Adam Smith'.
14. <http://adb.anu.edu.au/biography/wentworth-william-charles-2782> (accessed 8 January 2017).
15. Julie McIntyre and Jude Conway, 'Intimate, Imperial, Intergenerational: Settler Women's Mobilities and Gender Politics in Newcastle and the Hunter Valley', *Journal of Australian Colonial History*, vol. 19, 2017, pp. 161–84.
16. <http://adb.anu.edu.au/biography/windeyer-archibald-1055> (accessed 7 December 2017).
17. *The Colonist*, 2 March 1839.
18. Barrie Dyster, 'The 1840s Depression Revisited', *Australian Historical Studies*, vol. 25, no. 101, 1993, pp. 589–607.
19. Jack Sullivan, *Dr Henry Lindeman and Cawarra, Gresford, Part I*, Paterson Valley Historical Society, Paterson, 2014.
20. *Maitland Mercury*, 19 September 1891.
21. <www.jenwilletts.com> (accessed 7 December 2017).
22. Ludwig Leichhardt, 12 February 1843, in Geraldine Mate and Tracy Ryan (eds), *The Leichhardt Diaries: Early travels in Australia during 1842–1844, Memoirs of Queensland Museum Volume 7, Part 1*, Queensland Museum, South Brisbane, 2013, p.137.
23. Leichhardt, 18 December 1842, *The Leichhardt Diaries*, p. 86.
24. Leichhardt, *The Leichhardt Diaries*, p. 91.
25. Leichhardt, *The Leichhardt Diaries*, p. 91.
26. Leichhardt, *The Leichhardt Diaries*, p. 104.
27. Ludwig Leichhardt to Robert Lynd, 19 February 1843, in Leichhardt, *The Leichhardt Diaries*, p. 155.
28. Leichhardt, *The Leichhardt Diaries*, p. 154.
29. Leichhardt, *The Leichhardt Diaries*, p. 154.
30. New South Wales Legislative Council, Select Committee on the Condition of the Aborigines, *Replies to a Circular Letter, addressed to the Clergy, of all Denominations*, Printed by WW Davies, Government Printer, Sydney, 1846.
31. Cyrus Redding, *A History and Description of Modern Wines*, Whittaker, Treacher & Arnot, London, 1833, p. vii.
32. *Shipping Gazette and Sydney General Trade List*, 6 June 1846, p. 162.

Chapter 4 Creating an industry through co-operation: the Hunter River Vineyard Association, 1847–76

1. McIntyre, *First Vintage*, pp. 159–61.
2. *Maitland Mercury*, 1 April 1843.
3. *Maitland Mercury*, 8 April 1843.
4. James Belich, *Replenishing the Earth: The Settler Revolution and the Rise of the Anglo-world*, Oxford University Press, Oxford, 2009.
5. McIntyre, *First Vintage*, pp. 72, 226.
6. *Maitland Mercury*, 10 April 1847.
7. *Maitland Mercury*, 10 April 1847. See also Driscoll, *Beginnings*.
8. *Maitland Mercury*, 6 November 1847.
9. James King to Lord Grey, Secretary of State for the Colonies, 26 September 1849, in AA Company papers [78/1/19], NBAC, p. 425.
10. *Maitland Mercury*, 6 May 1848.
11. *Sydney Morning Herald*, 13 June 1849.
12. *Maitland Mercury*, 4 May 1850.
13. *Maitland Mercury*, 4 May 1850.
14. Earlier attributed to King, McIntyre, *First Vintage*, pp. 164–65.
15. *Maitland Mercury*, 9 November 1850.
16. *Maitland Mercury*, 17 August 1850.
17. William Windeyer to Maria Windeyer, 11 February 1851, Windeyer Family Papers, ML MSS 186/7, SLNSW.
18. *Maitland Mercury*, 28 August 1850.
19. William Windeyer to Maria Windeyer, 31 January 1850, ML MSS 186/7, SLNSW.
20. Didier Joubert to Maria Windeyer, 25 October 1849, ML MSS 186/7, SLNSW.

21 Merchant Rodd to Maria Windeyer, 25 February 1850, ML MSS 186/7, SLNSW.
22 *Maitland Mercury*, 11 May 1850.
23 William Windeyer to Maria Windeyer, 17 May 1851, ML MSS 186/7, SLNSW.
24 Jack Sullivan, *Patch and Glennie of Orindinna, Gresford: Early Wine Growing in the Paterson Valley, Part I*, Paterson Historical Society, Paterson, 2006.
25 James Simpson, *Creating Wine: The Emergence of a World Industry, 1840–1914*, Princeton University Press, Princeton, 2011, pp. 220–39.
26 McIntyre, 'Adam Smith'.
27 Various letters addressed to John Wyndham in Dalwood Papers, 1868–1902, ML MSS 8051, SLNSW.
28 <http://adb.anu.edu.au/biography/keene-william-3931> (accessed 17 June 2017).
29 *Maitland Mercury*, 6 May 1865.
30 *Maitland Mercury*, 6 May 1865.
31 *Maitland Mercury*, 6 May 1865.
32 *Leader* (Melbourne), 1 November 1884.
33 Lisa and Allan Thomas, 'Early Lochinvar and the North British Australian Loan Company', *Bulletin of Maitland and District Historical Society*, vol. 20, no. 2, 2013, pp. 9–18.
34 Republished from *Maitland Mercury*, *Empire* (Sydney), 3 May 1867.
35 McIntyre, *First Vintage*, pp. 121–24.
36 Jim Fitz-Gerald interview with Julie McIntyre, 14 April 2015, Vines, Wine & Identity Oral History Series, University of Newcastle Cultural Collections (UONCC).
37 *Sydney Morning Herald*, 1 August 1928.
38 Jim Fitz-Gerald, interview, 14 April 2015.
39 *Maitland Mercury*, 8 July 1875.
40 Jillian Oppenheimer, *Munros' Luck: From Scotland to Keera, Weebollabolla, Boombah and Ross Roy*, Ohio Productions, Walcha NSW, 1998, pp. 14–17.
41 Oppenheimer, *Munros' Luck*, pp. 19–20.
42 Hugh Wyndham to John Wyndham, 11 January 1869, Dalwood Vineyard Records 1868–1902, ML MSS 8051, Box 2/2, SLNSW.
43 John Glennie to John Wyndham, 17 May 1869, ML MSS 8051, Box 2/2, SLNSW.
44 See Robert Druitt, *Report on the Cheap Wines from France, Germany, Italy, Austria, Greece, Hungary and Australia: Their use in diet and medicine*, Second edition, Printed by Henry Renshaw, London, 1873.
45 CA Clements to John Wyndham, 17 April 1871, ML MSS 8051, Box 1/6, SLNSW.
46 See Stephen Garton, 'Once a Drunkard Always a Drunkard: Social Reform and the Problem of Habitual Drunkenness in Australia, 1880–1914', *Labour History*, vol. 53, 1987, pp. 38–53.
47 Druitt, *Report*.
48 Druitt, *Report*, from pp. iii–6.
49 Druitt, *Report*, p. 62.
50 Druitt, *Report*, p. 154.
51 These details are gathered from various sources, including John BS Hungerford, 'Audrey Wilkinson Wines', *Journal of the Hungerford and Associated Families Society* vol. 10, no. 3, 2010, p. 12.
52 <http://adb.anu.edu.au/biography/stephen-george-milner-1294/text3771> (accessed 14 December 2017).
53 Various entries for 1895, Audrey Wilkinson diaries, ML MS 532 Box 1(7), SLNSW.
54 Alexander Henderson, *The History of Ancient and Modern Wines*, Printed by Baldwin, Cradock and Joy, London, 1824, p. 1.
55 Author not stated, *Newcastle and Hunter District Historical Society Journal*, January 1948, p. 51. Thanks to Maitland District Historical Society for references for the Northumberland Hotel.
56 *Maitland Mercury*, 3 December 1872.
57 *Maitland Mercury*, 6 July 1874.
58 *Maitland Mercury*, 11 March 1876.
59 *Maitland Mercury*, 11 March 1876.
60 *Maitland Mercury*, 11 March 1876.
61 *Maitland Mercury*, 6 November 1847.

Chapter 5   A common people's paradise: German immigration from 1849

1 Saranna Scott to Helenus Scott, 30 September 1855, Letter No 28, Scott Family Papers, 1833–1964, ML MAV/FM3/69, SLNSW. Thanks to Mark Dunn.
2 See also Patrica Cloos and Jurgen Tampke, *Greetings from the Land where the Milk and Honey Flows: The German Emigration to New South Wales 1838–1858*, Southern Highlands Publishers, Canberra, 1993, p. 14. Note that there are many errors of fact and translation in this volume, according to Jenny Paterson, Review of Cloos and Tampke, *Greetings*, in *Descent*, December 1993, pp. 160–63.
3 McIntyre, *First Vintage*, p. 75.
4 Cloos and Tampke, *Greetings*, pp. 1–7.
5 Correspondence from Jenny Paterson, 10 December 2017.
6 *Maitland Mercury*, 14 August 1847.
7 Cited in Cloos and Tampke, *Greetings*, pp. 8–9, corrected in correspondence from Jenny Paterson, 13 November 2017.
8 Jenny Paterson, *The Ancestral Searcher*, vol. 20, no. 2, 1997.
9 Cloos and Tampke, *Greetings*, p. 14.
10 Correspondence from Jenny Paterson, 17 November 2017 and her articles on the 23 ships bringing these assisted German families to New South Wales published in *Ances-tree* (Burwood & District Family History Group), from 2003 to 2012.
11 *Maitland Mercury*, 14 April 1849.

12. Correspondence from Jenny Paterson, 17 November 2017.
13. Correspondence from Jenny Paterson, 11 December 2017.
14. Josef Horadam to his brothers and all close friends, 12 August 1849, Cloos and Tampke, *Greetings*, p. 90.
15. Horadam to his brothers and friends, 12 August 1849.
16. Friedrich Gerstäcker, in Peter Monteath (ed. and transl.) [and a translation team], *Australia: A German Traveller in the Age of Gold*, Wakefield Press, Adelaide, 2016, p. 24.
17. Gerstäcker, *Australia*, pp. 26–27.
18. Correspondence from Jenny Paterson, 17 November 2017.
19. Saranna Scott to Helenus Scott (Sydney) 27 May 1855, Glendon Letter No.7, Scott Family Papers, ML MAV/FM3/69, SLNSW.
20. See McIntyre and Conway, 'Intimate, Imperial, Intergenerational'.
21. Saranna Scott to Helenus Scott, 27 May 1855, Glendon Letter No.7, Scott Family Papers, ML MAV/FM3/69, SLNSW.
22. Saranna Scott to Helenus Scott, 3 June 1855, Glendon Letter No. 8, Scott Family Papers, ML MAV/FM3/69, SLNSW.
23. See for example Saranna Scott to Helenus Scott, 3 June 1855, Scott Family Papers, ML MAV/FM3/69, SLNSW.
24. Michael Jesser to Jakob Schmidt, 25 May 1853, in Marlene Buechele, 'Letters written by Wuerttemberg immigrants to New South Wales and addressed to their relatives and friends in the old country', *Margin*, no. 20, 1988, pp. 17–19.
25. Christian Carl Krust to Johann Specht, 29 August 1855, in Buechele, 'Letters', pp. 4–5.
26. Michael Jesser to Jakob Schmidt, in Buechele, 'Letters', p. 18.
27. WJ Goold, 'Peter Crebert, Mayfield', *Newcastle & Hunter District Historical Society*, vol. 2, no. 11, 1948, pp. 161–64.
28. *Newcastle Chronicle*, 20 June 1874.
29. For example *Newcastle Morning Herald*, 18 July 1878.
30. *NSW Government Gazette*, 15 December 1972, p. 5155.
31. Persons on bounty ships to Sydney, Newcastle, and Moreton Bay (Board's Immigrant Lists), Series 5317, Reel 2463, [4/4927], State Archives and Records NSW; Evelyn Boyce, *A History of the Weismantel Family*, Weismantel Family Reunion Committee, Krambach NSW, 1986.
32. Correspondence from Jenny Paterson, 17 November 2017.
33. AA Company, Despatch 23, Newcastle, 14 February 1863, Annual Report of Land Department, Port Stephens [yearly report printed in 1866], NBAC.
34. AA Company, Despatch 67, Newcastle, 12 February 1865, Annual Report of Land Department, Port Stephens, NBAC.
35. *NSW Government Gazette*, 9 May 1871, p. 1005; *NSW Government Gazette*, 8 May 1874, p. 1404.
36. Colin and Leah Weismantel, interview with Julie McIntyre, 7 December 2015, Vines, Wine & Identity Oral History Series, UONCC.
37. *Maitland Mercury*, 18 November 1865.
38. Correspondence from Jenny Paterson, 17 November 2017.
39. Information from Ken Knight, received via correspondence with Jenny Paterson, 17 November 2017.
40. *Sydney Mail*, 5 September 1891.
41. Correspondence from Jenny Paterson, 17 November 2017.
42. Correspondence from Jenny Paterson, 7 December 2017.
43. Gerhard Fischer, *Enemy Aliens: Internment and the Homefront Experience in Australia 1914–1920*, University of Queensland Press, St Lucia, 1989, p. 19. On some of the lasting effects of this internment period see Jillian Barnes and Julie McIntyre, 'A "Funny Place" for a Prison: Coastal Beauty, Tourism, and Interpreting the Complex Dualities of Trial Bay Gaol, Australia', in JZ Wilson, S Hodgkinson, J Piché and K Walby (eds), *The Palgrave Handbook of Prison Tourism*, Palgrave Macmillan, London, 2017, pp. 55–83.
44. Fischer, *Enemy Aliens*, pp. 17–18.
45. McIntyre, *First Vintage*, p. 76.
46. John Dunmore Lang quote from Archibald Gilchrist, *John Dunmore Lang: Chiefly autobiographical 1799–1878*, Jedgarm Publications, Melbourne, 1951, p. 496. Notes from Georg [sic] Schmid, Jenny's Paterson's Big Index, version printed 27 November 2017.
47. A. Cameron Archer, *A History of the Paterson Valley*, Paterson Historical Association, Paterson, 2019, forthcoming.

Chapter 6 New worlds of wine, at home and abroad: first wave globalists, 1877–1900

1. <http://adb.anu.edu.au/biography/james-john-stanley-3848/text6113> (accessed online 19 September 2017).
2. John Stanley James, in Michael Cannon (ed.), *The Vagabond Papers*, Melbourne University Press, Carlton, 1969, p. 158.
3. Kym Anderson, *The World's Wine Markets: Globalization at Work*, Edward Elgar, Cheltenham, 2004; Simpson, *Creating Wine*.
4. Kym Anderson and Vicente Pinilla (eds), *Wine Globalization: A New Comparative History*, Cambridge University Press, Cambridge, 2018.
5. James, *Vagabond*, p.158.
6. James, *Vagabond*, pp. 163–67.
7. James, *Vagabond,* p. 164.
8. Boris Sokoloff, *Aborigines in the Paterson Gresford Districts: Effects of Settlement*, Paterson Historical Society, Paterson, 2006, p. 25.
9. Lawrence Perry, '"Mission Impossible": Aboriginal survival before, during and after the Aboriginal Protection Era', PhD thesis, Adult Education, The Wollotuka Institute, University of Newcastle, 2014, see for example pp. 110–11.
10. *NSW Government Gazette*, 6 July 1898, p. 5057.

11. Heather Goodall, *Invasion to Embassy: Land in Aboriginal politics in New South Wales, 1770–1972*, Second edition, Sydney University Press, Sydney, 2005, pp. xxv, 80, 116–17, 145, 218.
12. McIntyre, *First Vintage*, pp. 123–31
13. John Strachan, 'The Colonial Identity of Wine: The Leakey Affair and the Franco-Algerian Order of Things', *The Social History of Alcohol and Drugs: An Interdisciplinary Journal*, vol. 21, no. 2, 2007, pp. 118–37.
14. Marion Fourcade, 'The Vile and the Noble: On the Relation between Natural and Social Classifications in the French Wine World', *The Sociological Quarterly*, vol. 53, 2012, pp. 524–45.
15. Richard Twopeny, *Town Life in Australia*, Printed by Elliot Stock, London, 1883, p. 67.
16. McIntyre, *First Vintage*, pp. 165–66.
17. (Henry Bonnard), *Report of the Executive Secretary on the Bordeaux International Exhibition of Wines*, 1882 Government Printer, Sydney, 1884, p. 6.
18. On Burgoyne see Simpson, *Creating Wine*, p. 269.
19. Diary of Arthur H Nicholson, wine buyer for PB Burgoyne & Co Melbourne from 1888 to 1903, MS 2128, SLSA.
20. McIntyre, *First Vintage*, p. 212.
21. Number of Vignerons (from Statistical Register), *Agricultural Gazette of NSW* 1892, pp. 144–47.
22. *Maitland Weekly Mercury*, 5 March 1898.
23. Michele Blunno, Report on the 1898 Vintage, *Agricultural Gazette of NSW*, vol. 8, p. 502.
24. Jakob B Madsen, 'Australian economic growth and its drivers since European settlement', in Simon Ville and Glenn Withers (eds), *The Cambridge Economic History of Australia*, Cambridge University Press, Cambridge, 2014, p. 32.
25. *Maitland Mercury*, 7 March 1891.
26. *NSW Government Gazette*, 24 December 1891, p. 10079.
27. *NSW Government Gazette*, 28 February 1895, p. 1.
28. *NSW Government Gazette supplement*, 7 April 1896, p. 2520.
29. Various entries for 1895, Audrey Wilkinson diaries, ML MS 532 Box 1(7), SLNSW.
30. *Newcastle Morning Herald and Miners' Advocate*, 7 August 1936; *The Cessnock Eagle and South Maitland Recorder*, 6 August 1940. Also detail in previous paragraph: *Maitland Weekly Mercury*, 19 March 1898.
31. <https://www.measuringworth.com/australiacompare/relativevalue.php> (accessed 26 January 2017).
32. Daybook, 1889–1905 and Cashbook, 1899–1904, Papers of John Murray Macdonald, Capricornia Central Queensland (CCQ) MS A2/2-17, Box 2, Central Queensland University Library.
33. *The Farmer and Settler*, 24 November 1911.
34. *Maitland Mercury Daily Mercury*, 29 December 1911.
35. *Maitland Daily Mercury*, 3 April 1894.
36. Obituary, *Maitland Daily Mercury*, 7 October 1938.
37. *Statistical Register of New South Wales, for the year 1888*, Government Printer, Sydney, 1889, pp. 244–45.
38. *Statistical Register for 1889*, pp. 296–97.

Chapter 7 Amid turmoil and temperance: the forgotten generation and the stoics, 1901–54

1. *Maitland Weekly Mercury*, 19 May 1900 (reprinted from an earlier report, as the vintage occurred in February).
2. For discussion of the first wave of wine globalisation ending with World War I, see Simpson, *Creating Wine*.
3. Nicholas Faith, *Liquid Gold, the Story of Australian Wine and its Makers*, PanMacmillan, Sydney, 2002, pp. 74 and 105; David Dunstan, *Better than Pommard!: A History of Wine in Victoria*, Australian Scholarly Publishing and the Museum of Victoria, Melbourne, 1994, p. 204.
4. The description of wine styles and classes is taken from Gunn and Gollan, *Report*, p. 21.
5. Dunstan, *Better than Pommard!*, pp. xvii, 158–60.
6. *Glen Innes Examiner and General Advertiser*, 25 October 1901.
7. *The Sydney Mail and New South Wales Advertiser*, 13 July 1901.
8. *Australian Town and Country Journal*, 10 August 1901, p. 4.
9. *Maitland Daily Mercury*, 21 August 1901.
10. *The Maitland Weekly Mercury*, 28 September 1901.
11. *Maitland Daily Mercury*, 26 September 1901.
12. <http://adb.anu.edu.au/biography/wilkinson-audrey-harold-9101> (accessed 17 October 2017); *Singleton Argus*, 25 July 1903.
13. *Sydney Morning Herald*, 21 December 1905.
14. *Sydney Morning Herald*, 15 June 1907.
15. *Sydney Morning Herald*, 22 October 1908.
16. *Sydney Morning Herald*, 7 April 1909.
17. *Freeman's Journal*, 24 March 1910.
18. *Wellington Times*, 27 April 1911.
19. *Sydney Morning Herald*, 5 June 1912.
20. *The Newsletter: an Australian Paper for Australian People*, 8 November 1913.
21. *Freeman's Journal*, 11 December 1913.
22. *The Cessnock Eagle and South Maitland Recorder*, 10 April 1914, see for example, *The Sun*, 5 June 1914.
23. *Sunday Times*, 8 November 1914.
24. Grateful thanks to Meredith Lake, author of *The Bible in Australia*, NewSouth, Sydney, 2018, for sharing her view that the window expresses family devotion, rather than an imperial theology.

25 Michele Blunno, 'Viticulture' in *New South Wales, the Mother State of Australia: A Guide for Immigrants and Settlers*, Issued by the Intelligence Department, Government Printer, Sydney, 1906, p. 250.
26 Blunno, *New South Wales*, p. 252.
27 *The Daily Telegraph*, 17 October 1913.
28 PJ Mylrea, *In the Service of Agriculture: A Centennial History of the New South Wales Department of Agriculture 1890–1990*, NSW Agriculture and Fisheries, Sydney, 1990, p. 59.
29 *Maitland Daily Mercury*, 15 January 1921.
30 *Maitland Weekly Mercury*, 16 October 1926.
31 *Sydney Morning Herald*, 21 April 1922.
32 *Newcastle Morning Herald and Miners' Advocate*, 5 April 1916, p. 5; *Newcastle Morning Herald and Miners' Advocate*, 22 July 1916, p. 4.
33 *NSW Government Gazette*, 22 December 1916, no. 223, 7683; *Sydney Morning Herald*, 22 July 1916.
34 *Report of the Royal Commission on Liquor Laws in New South Wales*, Government Printer, Sydney, 1954, p. 33, delivered by the commissioner, Justice Allan Maxwell (*1954 (Maxwell) Report*).
35 *The Land*, 13 July 1917.
36 Nadine Helmi and Gerhard Fischer, *The Enemy at Home, German Internees in World War I Australia*, UNSW Press, Sydney, 2011; Barnes and McIntyre, 'A "Funny Place" for a Prison'.
37 Liquor (Amendment) Act, George V, No. 42, 1919.
38 Liquor (Amendment) Act, 1923 (Act No. 51).
39 *Cessnock Eagle and South Maitland Recorder*, 29 June 1923.
40 *The Maitland Weekly Mercury*, 4 April 1925.
41 Gunn and Gollan, *Report*, pp. 56–59.
42 Maggie Brady, *Teaching 'Proper' Drinking? Clubs and Pubs in Indigenous Australia*, ANU Press, Canberra, 2018.
43 'Localities for the special pursuits shewn hereunder are indicated by green edging', compiled to 25 April 1919, National Library of Australia, FERG/3596, cited Bruce Scates and Melanie Oppenheimer, *The Last Battle: Soldier Settlement in Australia 1916–1939*, Cambridge University Press, Melbourne, 2016, (colour illustrations, n.p.).
44 Hector Trevena interviewed by Lindsay Francis, 26 June 2000, Treading Out the Vintage, OH 692/160, SLSA.
45 *Newcastle Herald*, 4 February 2014.
46 *Singleton Argus*, 19 November 1927.
47 Gunn and Gollan, *Report*, p. 3.
48 *Murrumbidgee Irrigator*, 12 June 1925.
49 *Report on Losses due to Soldier Settlement*, delivered by Justice Pike, Government Printer, Canberra, 1929; Scates and Oppenheimer, *Last Battle*, pp. 232–37.
50 Bill Gammage, cited in Scates and Oppenheimer, *Last Battle*, p. 245.
51 Various documents in the collection, Papers of Maurice O'Shea, Local Studies, Newcastle Region Library.
52 *Newcastle Morning Herald and Miners' Advocate*, 20 October 1924.
53 Jeans, *An Historical Geography*, p. 13.
54 See also McIntyre, Mitchell, Boyle and Ryan, 'We Used to Get and Give'; Julie McIntyre, 'Resisting Ages-old Fixity as a Factor in Wine Quality: Colonial wine tours and Australia's early wine industry as a product of movement from place to place', *LOCALE: the Australasian–Pacific Journal of Regional Food Studies*, vol. 1, no. 1, 2011, pp. 1–19.
55 *The Maitland Weekly Mercury*, 16 October 1926.
56 Christopher Fibbens, *Wood of the Cranes: The Story of St Mary on Allyn and Caergwrle, Allynbrook and the gentlefolk of the district*, Anglican Parish of Gresford & Paterson St Mary on Allyn Restoration Fund, Gresford, 2005, p. 45.
57 Fibbens, *Wood of the Cranes*, p. 46.
58 Gunn and Gollan, *Report*, p. 10.
59 Gunn and Gollan, *Report*, p. 10.
60 *Sydney Morning Herald*, 16 September 1931.
61 *Commonwealth of Australia Gazette*, vol. 35, 30 April 1931, p. 725.
62 *Smith's Weekly*, 25 June 1938.
63 *Newcastle Morning Herald and Miner's Advocate*, 10 February 1940.
64 *Newcastle Morning Herald and Miner's Advocate*, 10 February 1940.
65 *The Maitland Weekly Mercury*, 21 December 1901.
66 James Halliday, 'Halliday, Hungerford and the Hunter', *Journal of the Hungerford and Associated Families Society*, vol. 10, no. 3, 2010, p. 2.
67 *Newcastle Morning Herald and Miners' Advocate*, 16 January 1947.
68 *Newcastle Morning Herald and Miners' Advocate*, 17 February 1947.
69 Halliday, 'Halliday, Hungerford and the Hunter', p. 3.
70 *Cessnock Eagle and South Maitland Recorder*, 1 October 1954.

Chapter 8 Civilised drinking in a wine nation: the Renaissance generation, 1955–83

1 Karl Stockhausen, interview with Lindsay Francis, 29 June 2000, Treading Out the Vintage, OH 692/155, SLSA.
2 McIntyre and Germov, 'Who Wants to be a Millionaire?', pp. 65–84.
3 Eric Rolls, *A Celebration of Food and Wine, Of Grain, Of Grape, Of Gethsemane*, University of Queensland Press, St Lucia, 1997, p. 104.
4 Janis Wilton and Richard Bosworth, *Old Worlds and New Australia: The Post-war Migrant Experience*, Penguin, Melbourne, 1984, pp. 23–24.

5   Janet Semler, 'The Australian Wine Industry', *The Australian Quarterly*, vol. 35, no. 4, 1963, pp. 28–35.
6   Tanya Luckins, '"Pigs, Hogs and Aussie Blokes": The Emergence of the Term "Six O'Clock Swill"', *History Australia*, vol. 4, no. 1, 2007, pp. 8.1–8.17.
7   <http://adb.anu.edu.au/biography/maxwell-allan-victor-11091> (accessed 11 February 2018).
8   *1954 (Maxwell) Report*, p. 75.
9   McIntyre and Germov, 'Who Wants to be a Millionaire?'.
10 Michael Symons, *One Continuous Picnic: A Gastronomic History of Australia*, Melbourne University Press, Carlton [1982], 2007, pp. 252–53.
11 Karl Stockhausen, interview with Lindsay Francis.
12 Karl Stockhausen, interview with Julie McIntyre, 10 March 2015, Vines, Wine & Identity Oral History Series, UONCC.
13 André Simon, *The Wines, Vineyards and Vignerons of Australia*, Paul Hamlyn, London, 1967, p. 47.
14 On the inversion of beer and wine traditions, see Organisation for Economic Co-operation and Development (OECD), 'Alcohol Consumption', Factbook 2011–2012: Economic, Environmental and Social Statistics, OECD Publishing, <http://dx.doi.org/10.1787/factbook-2011-108-en> (accessed 15 May 2016).
15 Simon, *Wines*, p. 47.
16 Simon, *Wines*, p. 11.
17 Max Lake, 'The Wine', in WS Parkes, Jim Comerford and Max Lake, *Mines, Wines and People: A History of Greater Cessnock*, City of Greater Cessnock, Cessnock, 1979, p. 223.
18 Karl Stockhausen, interview with Julie McIntyre.
19 Max Drayton, interview with Rob Linn, 5 May 2003, Treading Out the Vintage, OH692/44, SLSA.
20 Philip Laffer, interview with Rob Linn, date not recorded, OH 692/83, SLSA.
21 See McIntyre and Germov, 'Who Wants to be a Millionaire?'.
22 Digby Matheson, interview with Rob Linn, Treading Out the Vintage, 5 May 2003, OH 692/104, SLSA.
23 Geoff Page, 'Wine buying: Pokolbin,' in Geoff Page, *The Question*, co-published with Philip Roberts, *Single Eye*, in *Two Poets*, University of Queensland Press, St Lucia, 1971, p. 36.
24 Phil Ryan, interview with Julie McIntyre, 28 May 2015 and 1 June 2015, Vines, Wine & Identity Oral History Series, UONCC.
25 Brian McGuigan, interview with Lindsay Francis, 27 Jun 2000, Treading Out the Vintage, OH 692/90, SLSA.
26 Ray Kidd, interview with Rob Linn, 6 May 2003, Treading Out the Vintage, OH 692/80, SLSA.
27 Len Evans, interview with Rob Linn, 6 May 2003, Treading Out the Vintage, OH 692/45, SLNSW.
28 Anderson et al, *Growths and Cycles*, p. 221.
29 Ivan Howard, interview with Julie McIntyre, 1 June 2015, Vines, Wine & Identity Oral History Series, UONCC.
30 Ivan Howard, interview with Julie McIntyre.
31 Facilities were based at Griffith, see Mylrea, *In the Service of Agriculture*, pp. 75–77.
32 Karl Stockhausen, interview with Julie McIntyre.
33 Brian McGuigan, interview with Lindsay Francis, OH 692/90, SLSA.
34 Brian McGuigan, interview with Lindsay Francis.
35 Keith Yore, interview with Julie McIntyre, 6 May 2015, Vines, Wine & Identity Oral History Series, UONCC.
36 Email correspondence from Christopher Barnes to Julie McIntyre, 14 March 2018.
37 Iain Riggs, interview with Julie McIntyre, 13 May 2015, Vines, Wine & Identity Oral History Series, UONCC.
38 Michael Foster, 'Wine', *The Canberra Times*, 13 June 1983, p. 11.

Appendix 1

Statistical problems and solutions

GIVEN THE RICH SEAM OF KNOWLEDGE OFFERED BY COLONIAL STATISTICAL data recorded from 1859 in the published Statistical Registers of New South Wales, we expected to encounter the figures on vine cultivation and winemaking as a stable scaffolding around which could be interlaced stories from letters, diaries and newspaper reports. Letters, and so on, are clouded by the politics of memory and freighted with the subjectivities of human hopes and dreams, and historians treat them somewhat warily. Instead, on working with the figures for Hunter winegrowing with statistician Paul Rippon, it became evident that from 1859 to 1900 the collection of numerical data was evolving and that the names of districts, for example, did not align during the whole period. Across these four decades, for instance, the names of parishes, counties and electorates change to reflect the growing population of certain centres. This makes it difficult (if not impossible) to line up the names of districts in the lettered record with the names of districts in the statistical record. Midway through this time period, the colonial statistician experimented with recording figures under different headings to explain geographical subregions, and then switched back to earlier methods. Moreover, when graphs are generated from the figures, there are anomalies that indicate the data may be flawed, and so we have left gaps in the graphs for these inconsistencies, to promote a focus on what may be gleaned from the figures. These graphs are still a window to the evolution of Hunter winegrowing in the crucial period of establishment of this region's wine industry.

APPENDIX 1   283

Australian, NSW and Hunter Valley wine production, 1859–1900

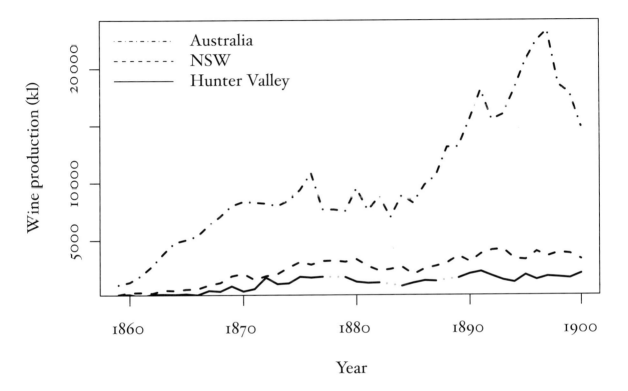

This shows how the Hunter's wine production compares to New South Wales production, and the wider national scale of wine output. Note that the gaps in the lines on the graph indicate where data is missing or ambiguous. The data was collected from Statistical Registers, and resources created by Kym Anderson. Graph provided by Paul Rippon.

Wine grape acres across regions, 1859–1900

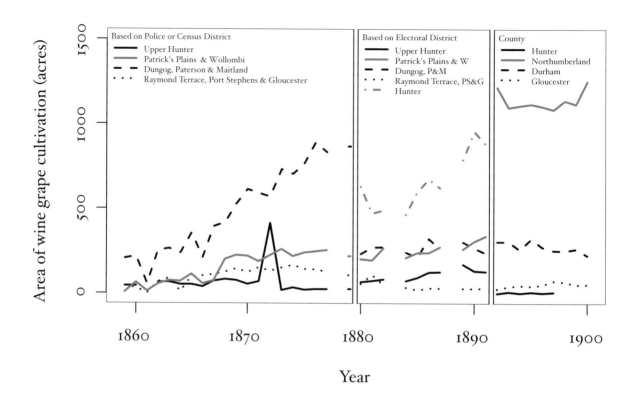

These graphs compare the vineyard acreages for wine grapes in the Hunter from 1859–1900 as they were reported in geographical categories. As the geographical categories changed during the period, it is not possible to align the figures over time. Places within the Hunter region are delineated with different terms for subregions in three distinct periods, and there are gaps and inconsistencies in the data. Nevertheless, this graph gives some idea of where plantings were occurring during the second half of the 19th century based on New South Wales statistical records. Graph provided by Paul Rippon.

APPENDIX 1   285

Comparative grape use in the Hunter Valley, 1859–1900

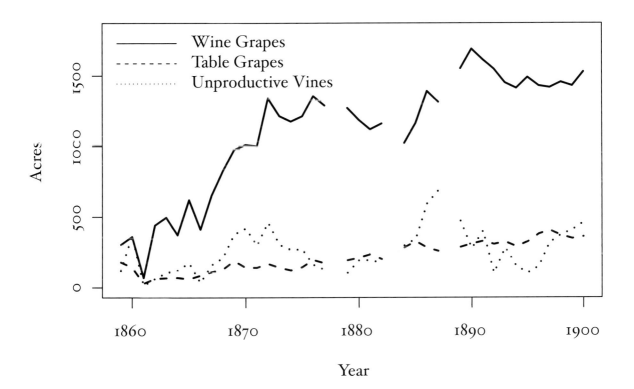

The dominance of wine grape vineyards in the Hunter, as opposed to table grapes, is a distinctive characteristic of the region's longevity. Note that gaps signal missing or ambiguous data. Graph provided by Paul Rippon.

Appendix 2

Cemeteries as sites of memory for the Hunter winegrowing community

MANY HUNTER CEMETERIES ARE THE FINAL RESTING PLACES OF MEMBERS of the region's winegrowing community. The National Trust of Australia recognises that

> As an expression of people's culture and identity, cemeteries comprise a fascinating resource which allow the community to delve back into their past. The monuments and graves represent the last public memorials of many people, both famous and unknown, who were intimately involved with the growth of the local area in which they are buried. In this way the headstones themselves, through the names, occupations, dates and epitaphs, provide a largely unique social, literary and economic record of the district (*National Trust Guidelines for Cemetery Conservation*, Second edition, NSW Department of Planning, Heritage Branch, Sydney, 2009, p. 3).

Some members of the Hunter winegrowing community are buried on private land, such as the Busby and Kelman family plot at Kirkton. However, most are in small public cemeteries throughout the region. This list has been compiled from contributions from Jim Fitz-Gerald, Pauline Tyrrell and several other members of the winegrowing community.

Branxton Cemetery
John Younie and Florence Tulloch

Campbell's Hill Cemetery, West Maitland
Thomas White Melville Winder and family (Windermere)
Emanuel Hungerford and some family

Dalwood Private Cemetery, Branxton
George Wyndham and family

Glenmore Cemetery
George Frederick McDonald and family (of Glenmore & Ben Ean)
Joseph and Mary Ann Drayton; various Drayton family members
Chick family members
Joass Family members

Kirkton Private Cemetery
William Kelman and family
John and Sarah Busby

St Anne's Anglican Cemetery, Gresford
Henry Lindeman
Arthur Lindeman

St Brigid Catholic Cemetery, Branxton
Perc McGuigan

St John's Anglican Cemetery, Clarence Town
Thomas and Miriam Holmes

St Peter's Anglican Cemetery, East Maitland
George Townshend

Our Lady of Lourdes Catholic Cemetery, Lochinvar
Catherine Terrier (wife of Philobert Terrier)

Pioneer Hill Cemetery, Raymond Terrace
Memorial for James King
Archibald Windeyer and John Windeyer

Tomago House Chapel, Tomago
Richard Windeyer

Wilderness Cemetery, Rothbury
Joseph Broadbent Holmes and family
Lambkin family members
Joass family members
Edward Tyrrell and his wife Susan Hungerford; EGY ('Dan') Tyrrell; Avery Tyrrell and his wife Dorothy Davey; Murray Tyrrell and his wife Ruth Church
Alfred Wilkinson, Frederick Albert Wilkinson and families, including Frederick's wife Florence Mary Hindmarsh Stephen, and her brother Alfred Farish Hindmarsh Stephen
George Campbell and his wife Mary; George Alexander Campbell; his son George Alexander Campbell and his wife Vera (vigneron of Daisy Hill)
Ed Jouault
Murray Robson

Bibliography

PRIMARY SOURCES

Archives & Selected Collections

Archives of the Royal Agricultural Society of New South Wales
British Library
Central Queensland University Library
    John Murray Macdonald Papers
Daniel Solander Library, Royal Botanic Gardens Sydney
Newcastle Region Library, Local Studies
    Maurice O'Shea Papers
Noel Butlin Archives Centre, Australian National University
    Australian Agricultural Company Records
    Lindeman's (Holdings) Ltd
State Archives and Records New South Wales
State Library of New South Wales
    Busby and Kelman Papers
    Dalwood Vineyard Records 1868–1902
    Dumaresq Papers
    Little Family Papers
    Scott Family Papers
    Wilkinson Family Papers
    Windeyer Family Papers
State Library of South Australia
    Treading Out the Vintage Oral History Series
University of Newcastle Cultural Collections
    Vines, Wine & Identity Oral History Series

Government inquiries

Report on Losses due to Soldier Settlement, Government Printer, Canberra, 1929, delivered by Justice Pike.

Gunn, J & Gollan, RM, *Report on the Wine Industry of Australia*, Government Printer, Canberra, 1931.

Report of the Royal Commission on Liquor Laws in New South Wales, Government Printer, Sydney, 1954, delivered by Justice Maxwell.

Publications

Atkinson, J, *An Account of Agriculture and Grazing in New South Wales and some of its most useful natural productions … as a result of several years residence and practical experience.* Printed for James Atkinson, London, 1826.

Blunno, M, 'Viticulture', in *New South Wales, the Mother State of Australia: A Guide for Immigrants and Settlers*, Issued by the Intelligence Department, Government Printer, Sydney, 1906.

Bonnard, H, *Report of the Executive Secretary on the Bordeaux International Exhibition of Wines*, 1882, Government Printer, Sydney, 1884.

Buechele, M, 'Letters written by Wuerttemberg immigrants to New South Wales and addressed to their relatives and friends in the old country', *Margin*, no. 20, 1988, pp. 17–19.

Busby, J, *A Treatise on the Culture of the Vine and the Art of Making Wine; compiled from the works of Chaptal, and other French writers; and from the notes of the compiler, during a residence in some of the wine provinces of France*, Printed by R Howe, Government Printer, Sydney, 1825.

— *A Manual of Plain Directions for Planting and Cultivating Vineyards, and for Making Wine in New South Wales*, Printed by R Mansfield, for the Executors of R Howe, Sydney, 1830.

— *Journal of a Tour through Some of the Vineyards of Spain and France* (1833), Facsimile edition, David Ell Press, Sydney, 1979.

Cloos, P & Tampke, J (eds), *Greetings from the Land where the Milk and Honey Flows: The German Emigration to New South Wales 1838–1858*, Southern Highlands Publishers, Canberra, 1993.

Cunningham, P, *Two Years in New South Wales: A series of letters comprising sketches of the actual state of society in that colony, of its peculiar advantages to emigrants, of its topography, natural history etc*, vol. 1, Printed by Henry Colburn, London, 1827.

Druitt, R, *Report on the Cheap Wines from France, Germany, Italy, Austria, Greece, Hungary and Australia: Their use in diet and medicine*, Second edition, Printed by Henry Renshaw, London, 1873.

Gerstäcker, F, *Australia: A German Traveller in the Age of Gold*, edited and translated by P Monteath [and a translation team], Wakefield Press, Adelaide, 2016.

James, JS, *The Vagabond Papers*, edited by M Cannon, Melbourne University Press, Carlton, 1969.

Jamison, J, *Report to Agricultural Society of New South Wales*, 23 February 1826, Sydney.

Henderson, A, *The History of Ancient and Modern Wines*, Printed by Baldwin, Cradock & Joy, London, 1824.

Hunter, C (ed), *The 1827 Newcastle Notebook and Letters of Lieutenant William S. Coke, H.M. 39th Regiment: Officer in charge of the military garrison stationed at Newcastle during 1827*, Hunter House, Raymond Terrace, 1997

Lake, M, 'The Wine', in WS Parkes, J Comerford & M Lake, *Mines, Wines and People: A History of Greater Cessnock*, City of Greater Cessnock, Cessnock, 1979, pp. 221–68.

Lang, JD, *A Historical and Statistical Account of New South Wales*, Cochrane & McCrone, London, 1834.

Mate, G & Ryan, T (eds), *The Leichhardt Diaries: Early travels in Australia during 1842–1844, Memoirs of Queensland Museum Volume 7, Part 1*, Queensland Museum, South Brisbane, 2013.

Monceau, MD, *Elements of Agriculture*, vol. 1, translated by Philip Miller, Printed for P Vaillant & T Durham, London, 1764.

Page, G, 'Wine Buying in Pokolbin', in Geoff Page, *The Question*, co-published with Philip Roberts, *Single Eye*, in *Two Poets*, University of Queensland Press, St Lucia, 1971.

Redding, C, *A History and Description of Modern Wines*, Whittaker, Treacher & Arnot, London, 1833.

Simon, A, *The Wines, Vineyards and Vignerons of Australia*, Paul Hamlyn, London, 1967.

Twopeny, R, *Town Life in Australia*, Printed by Elliot Stock, London, 1883.

Statutes

Liquor (Amendment) Act, George V, No. 42, 1919.

Liquor (Amendment) Act, 1923 (Act No. 51).

Government periodicals

Agricultural Gazette of New South Wales, Government Printer, Sydney, Years as noted.

Commonwealth of Australia Gazette, vol. 35, 30 April 1931.

NSW Government Gazette, Government Printer, Sydney, Years as noted.

Statistical Register of New South Wales, for the years 1859–1900, Government Printer, Sydney.

SECONDARY SOURCES

Publications

Abbott, JHM, *The Newcastle Packets and the Hunter Valley*, Currawong Publishing, Sydney, 1943.

Anderson, K, *The World's Wine Markets: Globalization at Work*, Edward Elgar, Cheltenham, 2004.

Anderson, K (with the assistance of N Aryal), *Growth and Cycles in Australia's Wine Industry: A Statistical Compendium, 1843 to 2013*, University of Adelaide Press, Adelaide, 2015.

Anderson, K & Pinilla, V (eds), *Wine Globalization: A New Comparative History*, Cambridge University Press, Cambridge, 2018.

Archer, AC, *A History of the Paterson Valley*, Paterson Historical Association, Paterson, 2019, forthcoming.

Barnes, J & McIntyre, J, 'A "Funny Place" for a Prison: Coastal Beauty, Tourism, and Interpreting the Complex Dualities of Trial Bay Gaol, Australia', in JZ Wilson, S Hodgkinson, J Piché & K Walby (eds), *The Palgrave Handbook of Prison Tourism* Palgrave Macmillan, London, 2017, pp. 55–83.

Belich, J, *Replenishing the Earth: The Settler Revolution and the Rise of the Anglo-world*, Oxford University Press, Oxford, 2009.

Blakeney, M, *The Protection of Geographical Indications: Law and Practice*, Edward Elgar, Cheltenham, 2014.

Boyce, E, *A History of the Weismantel Family*, Weismantel Family Reunion Committee, Krambach NSW, 1986.

Brady, M, *Teaching 'Proper' Drinking? Clubs and pubs in Indigenous Australia*, ANU Press, Canberra, 2018.

Brayshaw, H, *Aborigines of the Hunter Valley: A Study of Colonial Records*, Scone & Upper Hunter Historical Society, Scone, 1986.

Brodie, N, *1787*, Hardie Grant Books, Richmond Vic., 2016.

Campbell, JF, 'The Genesis of Rural Settlement on the Hunter', *Journal of the Royal Australian Historical Society*, vol. 12, no. 2, 1926, pp. 73–112.

Clarkson, C, Jacobs, Z ... Pardoe, C, 'Human Occupation of Northern Australia by 65,000 years ago, *Nature*, vol. 547, 2017, pp. 306–10.

Connor, J, *The Australian Frontier Wars 1788–1838*, UNSW Press, Sydney, 2002.

Demossier, M, *Burgundy: A Global Anthropology of Place and Taste*, Berghahn, Oxford, 2018.

Driscoll, WP, *The Beginnings of the Wine Industry in the Hunter Valley*, Newcastle History Monographs No. 5, Newcastle Public Library, The Council of the City of Newcastle, 1969.

Dunn, M, 'Aboriginal guides in the Hunter Valley, New South Wales', in T Shellam, M Nugent, S Konishi & A Cadzow (eds), *Brokers and Boundaries: Colonial exploration in Indigenous territory*, ANU Press, Canberra, 2016, pp. 61–84.

Dunstan, D, *Better than Pommard!: A History of Wine in Victoria*, Australian Scholarly Publishing and the Museum of Victoria, Melbourne, 1994.

Dunstan, D & McIntyre, J, 'Wine, olives, silk and fruits: The Mediterranean plant complex and agrarian visions for a "practical economic future" in colonial Australia', *Journal of Australian Colonial History*, vol. 16, 2013, pp. 29–50.

Dyster, B, 'The 1840s Depression Revisited', *Australian Historical Studies*, vol. 25, no. 101, 1993, pp. 589–607.

Faith, N, *Liquid Gold, the Story of Australian Wine and its Makers*, PanMacmillan, Sydney, 2002.

Fibbens, C, *Wood of the Cranes: The Story of St Mary on Allyn and Caergwrle, Allynbrook and the gentlefolk of the district*, Anglican Parish of Gresford & Paterson St Mary on Allyn Restoration Fund, Gresford, 2005.

Fischer, G, *Enemy Aliens: Internment and the Homefront Experience in Australia, 1914–1920*, University of Queensland Press, St Lucia, 1989.

Fourcade, M, 'The Vile and the Noble: On the Relation between Natural and Social Classifications in the French Wine World', *The Sociological Quarterly*, vol. 53, 2012, pp. 524–45.

Gammage, B, *The Biggest Estate on Earth: How Aborigines Made Australia*, Allen & Unwin, Sydney, 2011.

Garton, S, 'Once a Drunkard Always a Drunkard: Social Reform and the Problem of Habitual Drunkenness in Australia, 1880–1914', *Labour History*, vol. 53, 1987, pp. 38–53.

Gent, LC, *Gostwyck, Paterson 1823–2009*, Paterson Historical Society, Paterson, 2009.

Gilchrist, A, *John Dunmore Lang: Chiefly autobiographical 1799–1878*, Jedgarm Publications, Melbourne, 1951.

Goodall, H, *Invasion to Embassy: Land in Aboriginal politics in New South Wales, 1770–1972*, Second edition, Sydney University Press, Sydney, 2005.

Goold, WJ, 'Peter Crebert, Mayfield', *Newcastle & Hunter District Historical Society*, vol. 2, no. 11, 1948, pp. 161–64.

Grantham, I (ed.), *XYZ Goes North*, Convict Trail Project Occasional Monographs, Wirrimbirra Workshop, Kulnura NSW, 1999.

Halliday, J, 'Halliday, Hungerford and the Hunter', *Journal of the Hungerford and Associated Families Society*, vol. 10, no. 3, 2010: pp. 2–5.

Hancock, D, *Citizens of the World: London Merchants and the Integration of the British Atlantic Community, 1735–1785*, University of Michigan Press, Ann Arbor, 1998.

— *Oceans of Wine: Madeira and the Emergence of American Trade and Taste*, Yale University Press, Yale, 2009.

Helmi, N & Fischer, G, *The Enemy at Home, German Internees in World War I Australia*, UNSW Press, Sydney, 2011.

Hicks, B, 'Wollombi Landcare', in D Mahony & J Whitehead (eds), *The Way of the River, Environmental Perspectives on the Wollombi*, Wollombi Valley Landcare Group and The University of Newcastle, Newcastle, 1994.

Hungerford, JBS, 'Audrey Wilkinson Wines', *Journal of the Hungerford and Associated Families Society*, vol. 10, no. 3, 2010, p. 12.

Jeans, DN, *An Historical Geography of New South Wales to 1901*, Reed Education, Sydney, 1972.

Jupp, J & York, B, *Birthplaces of the Australian People: Colonial and Commonwealth Censuses, 1828–1991*, Centre for Immigration and Multicultural Studies, Canberra, 1995.

Karskens, G, *The Colony: A History of Early Sydney*, Allen & Unwin, Sydney, 2009.

— *People of the River*, Allen & Unwin, Sydney, forthcoming, ch. 9.

— 'The Settler Evolution: Space, Place and Memory in Early Colonial Australia', *Journal of the Association for the Study of Australian Literature*, vol. 13, no. 2, 2013, pp. 1–21.

Luckins, T, '"Pigs, Hogs and Aussie Blokes": The Emergence of the Term "Six O'Clock Swill"', *History Australia*, vol. 4, no. 1, 2007, pp. 8.1–8.17.

Ludington, C, *The Politics of Wine in Britain: A New Cultural History*, Palgrave Macmillan, London, 2013.

McGovern, PE, *Ancient Wine: The Search for the Origins of Viniculture*, Princeton University Press, Princeton, 2003.

McIntyre, J, 'Camden to Paris and London: The Role of the Macarthur Family in the Early NSW Wine Industry', *History Compass*, vol. 5, no. 2, 2007, pp. 427–38.

— '"Bannelong Sat Down to Dinner with Governor Phillip, and Drank his Wine and Coffee as Usual": Aborigines and Wine in Early New South Wales', *History Australia*, vol. 5, no. 2, 2008, pp. 39.1–39.14.

— 'Adam Smith and Faith in the Transformative Qualities of Wine in Colonial New South Wales', *Australian Historical Studies*, vol. 42, no. 2, 2011, pp. 194–211.

— 'Resisting ages-old fixity as a factor in wine quality: colonial wine tours and Australia's early wine industry as a product of movement from place to place', *LOCALE: the Australasian–Pacific Journal of Regional Food Studies*, vol. 1, no. 1, 2011, pp. 42–64.

— *First Vintage: Wine in Colonial New South Wales*, UNSW Press, Sydney, 2012.

— 'Trans-"imperial eyes" in British Atlantic Voyaging to Australia, 1787–1791', *History Australia*, 2018 (accepted for publication 23 December 2017).

— 'Wine Worlds are Animal Worlds too: Native Australian Animal Vine Feeders and Interspecies Relations in the Ecologies that Host Vineyards', in Nancy Cushing & Jodi Frawley (eds), *Animals Count: How population size matters in animal–human relations*, Routledge, London, 2018, forthcoming.

McIntyre, J & Conway, J, 'Intimate, Imperial, Intergenerational: Settler Women's Mobilities and Gender Politics in Newcastle and the Hunter Valley', *Journal of Australian Colonial History*, vol. 19, 2017, pp. 161–84.

McIntyre, J & Germov, J, 'Drinking History: Enjoying Wine in Early Colonial New South Wales', in S Eriksson, M Hastie & R Roberts (eds), *Eat History: Food and Drink in Australia and Beyond*, Cambridge Scholars Press, Melbourne, 2013, pp. 120–42.

— 'The Changing Global Taste for Wine: An Historical Sociological Perspective', in J Germov & L Williams (eds), *A Sociology of Food and Nutrition: The Social Appetite*, Oxford University Press, Melbourne, 2017, pp. 202–18.

— '"Who Wants to be a Millionaire? I do": Postwar Australian Wine, Gendered Culture and Class', *Journal of Australian Studies*, vol. 42, no. 1, 2018, pp. 65–84.

McIntyre, J, Mitchell, R, Boyle, B & Ryan, S, 'We Used to Get and Give a Lot of Help: Networking and Knowledge Flow in the Hunter Valley Wine Cluster', *Australian Economic History Review*, vol. 53, no. 3, 2013, pp. 1–21.

McManus, P, O'Neill, P, Loughran, R & Lescure, OR (eds), *Journeys: The Making of the Hunter Region*, Allen & Unwin, Sydney, 2000.

Madsen, JB, 'Australian economic growth and its drivers since European settlement', in S Ville & G Withers (eds), *The Cambridge Economic History of Australia*, Cambridge University Press, Cambridge, 2014, pp. 29–51.

Miller, J, *Koori, A Will to Win: the heroic resistance, survival & triumph of Black Australia*, Angus & Robertson, Sydney, 1985.

Mylrea, PJ, *In the Service of Agriculture: A Centennial History of the New South Wales Department of Agriculture 1890–1990*, NSW Agriculture and Fisheries, Sydney, 1990.

O'Connor, T, 'Charting New Waters with Old Patterns: Smugglers and Pirates at the Penal Station and Port of Newcastle 1804–1823', *Journal of Australian Colonial History*, vol. 19, 2017, pp. 17–42.

Oppenheimer, J, *Munros' Luck: From Scotland to Keera, Weebollabolla, Boombah and Ross Roy*, Ohio Productions, Walcha NSW, 1998, pp. 14–17.

Organization for Economic Co-operation and Development (OECD), 'Alcohol Consumption', *OECD Factbook 2011–2012: Economic, Environmental and Social Statistics*, OECD Publishing, Paris, <http://dx.doi.org/10.1787/factbook-2011-108-en> (accessed 15 May 2016).

Paterson, J, Review, Cloos & Tampke, *Greetings*, in *Descent*, December 1993, pp. 160–63.

Roberts, DA & Garland, D, 'The Forgotten Commandant: James Wallis and the Newcastle Penal Settlement, 1816–1818', *Australian Historical Studies*, vol. 41, no. 1, 2010, pp. 5–24.

Robertson, PS, *Proclaiming 'Unsearchable Riches': Newcastle and the Minority Evangelical Anglicans, 1788–1900*, Centre for the Study of Australian Christianity, Sydney, 1996.

Rolls, E, *A Celebration of Food and Wine, Of Grain, Of Grape, Of Gethsemane*, University of Queensland Press, St Lucia, 1997.

Scates, B & Oppenheimer, M, *The Last Battle: Soldier Settlement in Australia 1916–1939*, Cambridge University Press, Melbourne, 2016.

Semler, J, 'The Australian Wine Industry', *The Australian Quarterly*, vol. 35, no. 4, 1963, pp. 28–35.

Simon, A, *The Wines, Vineyards and Vignerons of Australia*, Paul Hamlyn, London, 1966.

Simpson, J, *Creating Wine: The Emergence of a World Industry, 1840–1914*, Princeton University Press, Princeton, 2011.

Sokoloff, B, *Aborigines in the Paterson–Gresford District: Effects of Settlement*, Paterson Historical Society, Paterson, 2006.

Sommers, BJ, *The Geography of Wine: How Landscapes, Cultures, Terroir, and the Weather Make a Good Drop*, Plume, New York, 2008.

Strachan, J, 'The Colonial Identity of Wine: The Leakey Affair and the Franco-Algerian Order of Things', *The Social History of Alcohol and Drugs: An Interdisciplinary Journal*, vol. 21, no. 2, 2007, pp. 118–37.

Sullivan, J, *George Townshend 1798–1872 and Trevallyn, Paterson River*, Paterson River Historical Society, Paterson, 1997.

— *Patch and Glennie of Orindinna, Gresford: Early Wine Growing in the Paterson Valley, Part I*, Paterson Historical Society, Paterson, 2006.

— *Dr Henry Lindeman and Cawarra, Gresford, Part I*, Paterson Valley Historical Society, Paterson, 2014.

Symons, M, *One Continuous Picnic: A Gastronomic History of Australia*, Melbourne University Press, Carlton [1982], 2007.

Thomas, L & A, 'Early Lochinvar and the North British Australian Loan Company', *Bulletin of Maitland and District Historical Society*, vol. 20, no. 2, 201, pp. 9–18.

Tobler, R, et al, 'Aboriginal Mitogenomes reveal 50,000 years of Regionalism in Australia', *Nature*, vol. 544, 2017, pp. 180–84.

Townsend, RD (ed.), *Rambles in New South Wales*, Sydney Grammar School Press, Sydney, 1996.

Walsh, B, *Voices from Tocal: Convict Life on a Rural Estate*, CB Alexander Foundation, Paterson, 2008.

— *European Settlement at Paterson River 1812 to 1822*, Paterson Historical Society, Paterson, 2012.

Werner, M & Zimmermann, B, 'Beyond Comparison: Histoire croisée and the challenge of reflexivity', *History and Theory*, vol. 45, February 2006, pp. 30–50.

White, RE, *Understanding Vineyard Soils*, Oxford University Press, Oxford, 2015.

Wilton, J & Bosworth, R, *Old Worlds and New Australia: The Post-war Migrant Experience*, Penguin, Melbourne, 1984.

Wood, WA, *Dawn in the Valley: The Story of Settlement in the Hunter River Valley to 1833*, Wentworth Books, Sydney, 1972.

Theses

Archer, AC, 'Social and Environmental Change as Determinants of Ecosystem Health: A case study of social ecological systems in the Paterson Valley NSW Australia', PhD thesis, Life and Environmental Sciences, University of Newcastle, 2007.

Dunn, M, 'A Valley in a Valley: Colonial struggles over land and resources in the Hunter Valley NSW 1820–1850', PhD thesis, History, University of NSW, 2015.

Honan, D, '"The Elements Are Against Us": Hunter Valley wine and the paradox of an enduring relationship between people and environment', Honours thesis, History, University of Newcastle, 2017.

McIntyre, JA, '"Civilized" Drink and a "Civilizing" Industry: wine growing and cultural imagining in colonial New South Wales', PhD thesis, History, University of Sydney, 2009.

Pemberton, P, 'The London Connection: the formation and history of the Australian Agricultural Company 1824–1834', PhD thesis, History, Australian National University, 1991.

Perry, L, '"Mission Impossible": Aboriginal survival before, during and after the Aboriginal Protection Era', PhD thesis, Adult Education, The Wollotuka Institute, University of Newcastle, 2014.

Walsh, B, 'Heartbreak and Hope, Deference and Defiance on the Yimmang: Tocal's convicts 1822–1840', PhD thesis, History, University of Newcastle, 2007.

Websites

A History of Aboriginal Sydney, < historyofaboriginalsydney.edu.au>

Australian Broadcasting Corporation, ABC Online, < abc.net.au>

Australian Dictionary of Biography, National Centre of Biography, Australian National University, Canberra, < adb.anu.edu.au>

Free Setter or Felon, <jenwillets.com>

National Library of Australia, Trove, < trove.nla.gov.au/>

Paterson River History, < patersonriver.com.au>

Relative Value, <https://www.measuringworth.com/australiacompare/relativevalue.php>

Pokolbin, c1949. Photograph by Max Dupain. Tulloch Family Collection, courtesy of J.Y. Tulloch.

Acknowledgments

MANY MINDS AND HANDS SHAPED THE HUNTER WINE REGION FROM 1828, through a continuous and intricate interweaving of habit into rituals and identities, so that it is no longer possible to imagine 'Hunter' without 'wine'. Many people have also shaped this book from its origins some years ago when fellow academics at the University of Newcastle, classicist Bernie Curran and historian Roger Markwick, introduced us to members of the (then) Hunter Valley Wine Industry Association. When we met the HVWIA Legends Sub-committee – Brian McGuigan, Jay Tulloch and Phil Ryan – in 2011, they were seeking new insight into the evolution of the Hunter wine community, and the places of importance in that emergence.

Up until this time, the wine industry used its past for branding, viewing 'history' as organisational, rather than tied to wider currents of scholarship. At the same time, we were participating in an international burgeoning of wine studies, looking from our fields of history and sociology at the workings of the wine world. Over time, our collaboration with Brian, Jay and Phil on small projects about Hunter wine history gave rise to a plan to draw together our different worlds in a major study of wine, history and the social behaviour of a wine community. In 2014, our collaboration with the (now) Hunter Valley Wine & Tourism Association, along with Newcastle Museum, received Australian Research Council (ARC) Linkage Project funding for a three-year program of inquiry (LP140100146), and the result is this book, a new series of oral history interviews online, and a Newcastle Museum exhibition.

Before now, many books within the wine world have considered 'Hunter wine'. Some of the best known that have focused on the region are Max Lake's *Hunter Wine* (Jacaranda Press, Brisbane, 1964) and James Halliday's *The Wines & History of the Hunter Valley* (McGraw-Hill, Sydney, 1979). Campbell Mattison has also applied his great powers of imagination in *Wine Hunter: The man who changed Australian Wine* (Hachette, Sydney, 2006). But 'history' in these books is at times at odds with the historical record, particularly Lake's *Hunter Wine*, and especially in relation to wrongly according the dates and detail of James Busby's role in Hunter and Australian winegrowing. We have taken a new approach to the Hunter wine past, by using archives and other historical artefacts to trace the origins and continuity of the Hunter wine community, and the evolution of the wine, within broader currents of change in wider worlds of society, economy and consumer taste. Our research has taken the best part of four years and engaged the guidance of many people, some of whom we thank here, and others to whom we expressed our gratitude as the project unfolded.

Contemporary members of the Hunter wine community have taken a large role in bringing this volume to fruition. In particular, Brian and Fay McGuigan, Jay and Julie Tulloch, Phil and Sylvia Ryan, and Bruce and Pauline Tyrrell. Also, Lindy Hyam, Garth Eather, Richard Hilder and Mike Wilson. Great appreciation is due to Newcastle Museum's former director Gavin Fry for his role in early project discussions, and to current Newcastle Museum director Julie Baird, a partner investigator on our ARC Linkage, whose peerless skill in bringing to life the stories of objects has been a cornerstone of the project.

Members of the wider community have also been of critical importance: Cameron and Jean Archer, Brian Walsh, Jim Fitz-Gerald, Jenny Paterson, Keri Negline, Kylie Rees, Heather McIntyre, Virginia Newell, Mark Burslem, Max Allen, Mike Wilson, Brian Miller and Maizie Denzin.

Thanks to our colleague Paul Rippon for statistical and graphing skills and to Melissa Rey-Lescure for illustration. And to other colleagues who have contributed in myriad, invaluable ways: Nancy Cushing, Grace Karskens, Richard Waterhouse, Mark Dunn, Lisa Murray, David Roberts, Lauren Williams, Catherine Oddie, Jenny Noble, Linda Hutchinson, Jarrod Skene, Jessie Reid, Gillean Shaw, Kathleen Brosnan, David Dunstan, James Simpson, Kym Anderson, Robin Callister, Kristi Street, Lyndall Ryan, Anne Bickford, Robert Mason, Celmara Pocock, Cathy Coleborne, Victoria Haskins, Kate Arrioti, Philip Dwyer, Libby Roberts-Petersen, Kit Candlin, Jodi Frawley, Lisa Featherstone, Jeffrey Pilcher, Maggie Brady, Jillian Barnes, Rumina Dhalla, Sidsel Grimstad, Sacha Davis, Barbara Santich, Jane Carr, Jaqueline Dutton, Peter Howland,

Corinne Marache and Marc de Ferriere. We could not have completed the project without the superlative efforts of students who worked closely on different aspects of the research: Cathy Brett, Christine Lawrie, Brian Roach, Mikaël Pierre, Don Seton Wilkinson and Daniel Honan. Mikaël, Don and Daniel are among a new generation of historians inquiring into the role of winegrowing in regional and transnational contexts.

Dozens of staff at our university library, and in public archives and libraries, have been instrumental in the finely grained process of finding material to bring together a rich collection of text, images, objects and memories, in particular Gionni di Gravio and Ann Hardy at the University of Newcastle Cultural Collections; Kelli Stidiford at Central Queensland University, Rockhampton campus; and Karen Finch at the Royal Agricultural Society of New South Wales. At the State Library of New South Wales, Richard Neville, Rachel Franks and Andy Carr have generously supported this project, most recently in regard to the Library Grape (see Conclusion). Also, Kirsten Clancy, Nicole Simons, Jan Pelosi and Christine Brett Vickers. And, at Newcastle Region Library's Local Studies, Suzie Gately, Alex Mills, Sue Ryan and Chloe O'Reilly. Thanks to Tocal Agricultural Centre library for their digitisation service. Acknowledgment must be given, as well, to the National Library of Australia for hosting the Trove database, without which contemporary digital humanities in Australia would be profoundly impoverished.

A good deal of the arguments herein were sharpened at conferences, seminar presentations and in our cited publications. These fora are a crucial part of our scholarly practice, and we are especially pleased to lead a new group of international researchers in wine studies in the humanities and social sciences. Our cross-disciplinarity requires us to convince our 'home' disciplines of history and sociology not only of our reasoned use of evidence, but also of the growing base for a studies field on wine. Through the Worlds in a Wine Glass Conference at the Menzies Centre for Australian Studies, King's College London in 2016 we began to create depth for this field. This meeting was achieved with the significant support of Menzies Centre director Simon Sleight.

We would also like to acknowledge the labour of the editors of journals and edited collections, and peer reviewers whose time and effort are crucial in the creation of new knowledge from the vast quantity of information increasingly available to researchers in the global, digital age. Our inquiries for this project led to several scholarly publications. These publications are cited in notes in this book and listed in the Bibliography. By referencing this scholarship we offer access to the outcomes of the wider ARC project. If we have done our job properly, some of these peer-reviewed publications will stimulate debate in our 'home' disciplines.

We are privileged to be published by NewSouth Publishing and are tremendously grateful to Elspeth Menzies for her confidence in our concept, and for her ever-elegant assurance of overseeing. Copy editor Diana Hill provided this book with the literary equivalent of fining wine after fermentation. Fining is one of the processes of removing unwanted substances to achieve a crystalline clarity of uninterrupted light through wine in a transparent glass. Diana patiently and deftly fostered the stories in this book into clearer view, a contribution that must not go unsung. And we recognise Emma Hutchinson's extraordinary forbearance and skill in guiding this book to its beautiful, orderly form.

Finally, thank you a thousand times to our families. As Julie has vastly more immediate relatives than John we have not enumerated them as individuals, but nonetheless extend our unceasing gratitude for graciously supporting all that we do.

Index

Page numbers in *italics* refer to illustrations. Introductory notes and appendices have not been indexed. Grape and wine names follow the capitalisation guidelines on pages vi–vii.

A

Aboriginal Australians *16*, *26*
   dispossession and violence 36, 44–45, 54, 64–65, 162–63
   interactions with settlers 69, 80–81, 98–100, 150–51, 206
   land management 15, 26, 30–31, 50, 91
   working on estates 45, 59, 62–63, 70–71
Account of Agriculture and Grazing in New South Wales 41–42
advertising 234, 244, 246–47, 249
Allandale, NSW *see* Norwood
Anderson, Kym 158, 252
Archer, Cameron 50, *157*
Ash Island, Newcastle, NSW 51, 139
Ashmans, Pokolbin, NSW 122–23, 180
Atkinson, James 41–42
Aucerot 123, 153, 180, 242
Australian Agricultural Company
   employees 59, 65, 81, 82, 152
   as winegrowers 2, 5, 43–44, 69, 83–86
   see also Tahlee
Australian Women's Weekly 234, *236–37*, 244, *256*
awards 113–14, 115, 154, 166–67, 198–99, 239 *see also* exhibitions

B

Bambach, Anton 137, 153–54
Barnes, Christopher 262
Bathurst, Henry, 3rd Earl 33, 44
Bebeah, Singleton, NSW 123–25, 166, 167, 170, *171*
Belich, James 104
Bellevue, Pokolbin, NSW 123, 179, *248* *see also* Drayton family

Ben Ean Moselle 235, 243–44
Ben Ean, Pokolbin, NSW *175*, 175–77, 181, 199, 238, 240, *245* *see also* John Murray Macdonald; Lindeman's Ltd
Bengalla, Muswellbrook, NSW 51, 63–64
Black Cluster *17*, 62, 63, 74, 82, 84–86, 110, 112
Black Damascus 60, 63
Black Hamburgh *17*, 74, 82, 83, 85–86
Blaxland, George 139
Blaxland, Gregory 5, 23, 70
Blunno, Michele 168, 176–77, 192, 201–202
Bonnard, Henry 167–68
boosterism 103–104, 274n22
Botanic Gardens, Sydney *12*
   source of vines 1, 8–9, 46, 63, 70, 73, 112, 160
   varieties of vines *10*, 17–18, 50, 60
Bourke, Richard 137–38
Boydell, Charles 59, 61–63, 109, 114, 149 *see also* Camyr Allyn
Boydell family 162, 214
Brady, Maggie 206
brandy 105, 153, 155, 206–207 *see also* fortified wines
Branxton, NSW 133, 173, 177–78, 222 *see also* Dalwood; Fern Hill
Brecht, Carl 137, 154–55, 177 *see also* Rosemount
Brecht, Will 137, 154–55, 269
Brisbane, Thomas 3, 7, 18, 36
Brix 11, 117 *see also* sugar
Brokenwood, Pokolbin, NSW 263–65
Browne, John 124
Bukkulla, Inverell, NSW 71, 125–26, 128, *129* *see also* John Wyndham
Bulletin, The 251
Burgundy 17–18, 62, 74, 97, 188, 199, 213, 215, 219
   Millers Burgundy *17*, 74
Burnett, William 81, 82, 113
Busby family 6, 71–74

Busby, James
   Journal 9, 73
   Manual 8–9, 47–48, 60, 64, 69, 83–84
   myths about 2–4, 9–11, 199
   Treatise 6, 51, 64, 131
   vines 8–11, 60, 70, 73–74, 112

C

Cabernet Sauvignon *17*, 82, 110, 241–42
Caergwrle, Paterson, NSW 63, 214
Caerphilly, Rothbury, NSW 122
Camyr Allyn, Gresford, NSW *26*, *29*, 61–63, 98, 151, *162* *see also* Charles Boydell
Cape colony, South Africa
   tours of Constantia 72, 82
   trade competition 44, 86, 108
   vines and wines 5, 7, 17–18, 43, 63–64
Capper, Edward 199 *see also* Catawba
Carmichael, Henry 87–88, 103, 105, 107, 109–11, 135, 139
Carmichael, John 79–80, 100
Caswell family 81
Caswell, William 71, 81, 90, 139
Catawba, Pokolbin, NSW 184, *185*, *186*, *189*, *190* *see also* Edward Capper
Cawarra, Gresford, NSW 92, 167, 180, 199, 200 *see also* Henry Lindeman; Lindeman's Ltd
cemeteries 200–201
Cessnock, NSW 29, 189, 196, 219, 242
Cessnock Vintage Festival 244
Chevron Hilton Hotel, Sydney 250–51
civility, and wine 8, 65, 88, 119, 128, 131
claret 6, 17, 82, 110, 175, 188, 199, 213, 215, 219
class
   and colonial society 8, 40–42, 65, 75–76, 88–89, 105
   and drinking culture 39, 55–56, 101, 116, 189, 204, 231–32, 234–35
   working class immigration 137–38, 143–44
Clevedon, Gresford, NSW 173

climate of Hunter 11, 148, 157, 211, 215, 241–42, 265–66
climate parity 43–44, 120
coal 28, 32, 33, 189, 215, 261
Connor, John 45
Constantia, Cape Colony 72, 82
Constantia (wine) 7, 18, 63
convict labour 4, 5
 and alcohol 191
 assigned to estates 46–47, 50, 54, 59, 75, 124
 end of 69, 105, 136
Coolalta, Pokolbin, NSW 167, 174, 199–200 see also Wilkinson family
coopers 138, 139, 142, 155
Corner, Stewart 196
corporate expansion 197–200, 222, 232, 234, 252, 254
Cory, Edward Gostwyck 49, 51, 54, 109
Cory, John 51, 54, 109
Craig, George 184–85
Crebert, Peter 151–52, 160
Cunningham, Peter 31, 41, 56, 57

D

Daisy Hill, Rothbury, NSW 168, 264
Dalwood, Branxton, NSW *134*
 Aboriginal Australians at 24, 98
 descriptions by visitors 94, 161, 170, 172
 establishment 48, 60, 69–71
 purchase of 182, 197, 199, 258
 wines 125–26, 133–34, 166–67, 174, 182, 221
Darling, Ralph 1, 44–45, 58
David Cohen & Co 125, 128
Davoren, John 182
De Beyer, Henry 179, 180
Denman, NSW *see* Merton; Pickering; Rosemount; Verona
Department of Agriculture 255–56
Despeissis, J Adrien 168
diseases *see* grape vines, diseases and pests
Doradillo 206–207
downy mildew 202–203, 207
Doyle family 120, 173
Doyle, James 167, 170, 173

Drayton family 123, 154, 160, 181, *248*, 249, 262
Drayton, Joseph 123
Drayton, Max 242–43
drinking culture 22–23, 38–39
 and class 39, 55–56, 101, 116, 189, 204, 231–32, 234–35
 and drunkenness 55–56, 86, 126, 191
 popularity of wine 12, 189, 230–31, 234–35, 240, 242–47, 249–51
 wine as civility 8, 65, 88, 119, 128, 131
 see also licensing laws; temperance
Driscoll, WP 23, 49
Druitt, Robert 126, 128
dry wines 11–12, 187–88, 196, 205, 215–17, 221, 240
Duhamel du Monceau, Henri-Louis 76
Dumaresq, Henry 48, 57–59
Dumaresq, William 14–15, 31, 45, 57–58
Dunmore, Maitland, NSW 54–55, 149–51
Dunstan, David 191–92
Dupain, Max 223, 225

E

economic depression 91, 177, 215, 261
Eelah, Maitland, NSW 153–54, 160, 172
Elliott, Bob 219
Elliott family 181, 207, 219, 252
emancipists 8, 33, 40, 76 *see also* class
equipment *see* wine making, equipment
espaliers *see* grape vines, cultivation
Evans, Len 250–52
exhibitions 92, 105, 165–68, 187, 198–99, 205 *see also* awards
exports 86, 101, 125, 159, 169, 174, 185, 197, 216

F

Fallon, James 125
Federation 23, 187, 188, 191–92, 197
Fern Hill, Branxton, NSW 125, 172
festivals 244–45
fire farming 30–31, 50 *see also* Aboriginal Australians, land management
Forsythe, Alec 262

fortified wines 11–12, 187–88, 213, 215, 221, 234
France
 Bordeaux exhibition *118*, 165–68
 vines 9, 18, 73–74, 163, 202
 experience from 6, 72–73, 92, 117, 119, 120, 168, 208
 wines from 6, 19, 88, 126, 165, 166, 182, 216
Francis, Lindsay 258
Fraser, Charles 1, 15, 42
free trade 87–88, 125, 191
 see also tariffs
Frontignac 17, 18, 60, 63, 188

G

Gammage, Bill 30, 207
Gent, Lesley 54
geography *see* Hunter region, geography and landscape
German immigrants 60, 146–55, 157, 191, 204
 bounty scheme 110, 136–39, 142–45, 155–56
German wine 39, 55, 96, 110, 145–46
Gerstäcker, Friedrich 145–46
Glendon, Singleton, NSW 51, *94*, 99
 Aboriginal Australians at 26, 98–99, 151
 German immigrants at 136, 139, 147–49
 wine making at 96–98
 see also Helenus Scott; Saranna Scott
Glen Elgin, Pokolbin, NSW 177–78, 185, 219, *233*, 293 *see also* John Tulloch; Hector Tulloch
Glennie, John 71, 114–15, 125, 161
globalisation *see* wine globalisation
Gostwyck, Paterson, NSW 51, *52*–53, 54, 133
Gouais 17, 50, 60, 74, 98
Gouget, Monsieur 182
grape varieties
 at Botanic Gardens *10*, 17–18, 50, 60
 identification of 37, 74, 112
 and wine names 7, 19, 39, 164–65, 243
 see also individual varieties by name

grape vines
    from Botanic Gardens 1, 8–9, 46, 63, 70, 73, 112, 160
    cultivation 62, 74–75, 76, 95–96, 154, 169, 170, 212–13
    diseases and pests 120–21, 163–64, 168, 201–203, 207
    exchange of cuttings 50–51, 60, 63–64, 112, 130–31
    imports to Australia 5, 9–10, 69, 73
    suitable soils 18–19, 42–43, 211, 215, 216, 242
    symbolism 39, 77, 80
    see also vineyards
Great Exhibition of 1851, London 110–11
Green, Walter 174–75
Gresford, NSW 27, 200 see also Camyr Allyn; Cawarra; Clevedon; Lewinsbrook; Orindinna;
Greta, NSW 177
    Migrant Camp 230, *233*

H

Habig, Peter and Regina 136, 149
Halliday, James 218–19, 264–65
Hardy, Thomas 187 see also Thomas Hardy & Sons
Harpers Hill, NSW 173
health benefits of wine 126, 128
Hely, Frederick 47–49, 60
Henderson, Alexander 131
Hermitage 133, 155, 170, 172, 173, 174, 180, 181, 213
    Red 17, 18, 74, 95, 96, 109, 110, 166, 169
    White 95 , 96, 110, 112, 166, 169, 215
Hickey, Edwin 92–93, 107, 110, 139
Hicks, Bill 30
historiography 22–24, 273n37
Hock 39, 110, 145–46, 175, 188, 192, 199, 215
Holmes, Charles Phillip 122, 196, 200
Holmes family 122, 168 see also The Wilderness
Holmes, Joseph Broadbent 121–22
Honan, Daniel 22
Horadam, Josef 144–45

Howard, Ivan 255–57, 262
HRVA see Hunter River Vineyard Association
Hunter region 1–2, 27–28, 236–37
    climate 11, 148, 157, 211, 215, 241–42, 265–66
    geography and landscape 1, 14–15, 18, 27–31, 50, 56, 160–61, 215, 216, 242
    opinions as wine region 11–12, 44, 104, 215–16, 264
Hunter River Agricultural Club 1, 3, 49, 68
Hunter River Agricultural Society 106, 132
Hunter River General Advertiser 103
Hunter River Vineyard Association (HRVA)
    meetings 108–11, 117, 119
    members 93, 106–107, 113, 114, 115, 128
    organisation 102, 106–108, 131–32, 135
Hunter Valley Vineyard Association (HVVA) 262–63
Hunter Vintage Festival 244–45, *253*
Hutcherson, AY 184, 196
HVVA see Hunter Valley Vineyard Association

I

Idstein, 'Nancy' 147–48
Institute of Masters of Wine 240
Inverell, NSW see Bukkulla
Invermien, Scone, NSW 49, 57
Irrawang, Raymond Terrace, NSW 80, *147*
    establishment 78–81, 172
    German immigrants at 139, 146–47
    wines from 2, 101, 145–46, 166
    see also James King
irrigation 206, 260–61
Ivanhoe, Pokolbin, NSW 167, 196, 270–71

J

James, John Stanley 158, 160–62
Jamison, John 44, 87
Jeans, DN 211

Jesser, Michael and Maria 149
Joubert, Didier 113
Journal of a Tour through Some of the Vineyards of Spain and France 9, 73

K

Kaludah, Lochinvar, NSW 119–20, 170
Karskens, Grace 30, 41
Keene, William 117, 119
Kelman, James 130, 166, 169
Kelman, William
    and Busby family 3, 71–73
    grapes and wine 48, 73–74, 91, 95–96, 110, 112
    role in Hunter wine 47, 71, 74–75, 135
    see also Kirkton
Kidd, Ray 219, 238, 243, 250
King, James
    role in Hunter wine 47, 79, 87, 90, 135
    scientific interests 78, 93, 110, 111
    wines 2, 90–91, 101, *101*, 109, 113, 166
    see also Irrawang
Kinross, Raymond Terrace, NSW 90, 112, 139, 173
Kirchner, Wilhelm 138–39, 142–43, 145
Kirkton, Singleton, NSW
    establishment 71–75
    German immigrants at 152
    purchase by Lindeman's 199–200
    vine plantings at 9, 11, 48, 73–74
    wine making at 95–96
    wines from 17, 110, 130, 166, 169–70
    see also William Kelman
Krust, Christian Carl 149–50

L

labour
    immigration of skilled workers 55, 69, 120, 136–39, 155–56
    seasonal 99, 133, 134–35, 215, 230
    shortages 104–105, 211, 220–21
    see also convict labour
Laffer, Philip 243
Lake, Max 241–42, 262
Lake's Folly, Pokolbin, NSW 241–42
Lambruscat 155, 166

land ownership 2, 17, 36, 40–41, 65, 91, 133, 137, 206–207, 252
Lang, Andrew 54–55, 81, 105, 107, 108, 109, 120, 157 *see also* Dunmore
Lang, John Dunmore 55–56, 69, 77, 87, 137
legislation 103, 104, 108, 116, 151, 196, 216 *see also* licensing laws
Leichhardt, Ludwig 93–100, 139
Lewinsbrook, Gresford, NSW 66–67, 68–69
licensing laws 69, 103, 116, 151–52, 153, 203–206, 217–19, 231
Liebig, Justus von 110, 111, 129
Lindeman, Henry 92, 111–12, 126, 128, 135, 167, 200
Lindeman's Ltd 199–200, 253, 257
    Sydney wine cellars 176, 197, 217, 218
    wine making at 238–39
    wines from *198*, 198–99, 243–44
    see also Cawarra; Ben Ean
Little, Francis 49, 57
Lochinvar, NSW 144, 170 *see also* Kaludah; St Helier's
Lowe, Nathaniel 44–45
Luskintyre, Singleton, NSW 88–89

M

Macarthur family 15, 55, 69, 88–89, 137–38, 144
Macarthur, John 2, 15, 18, 55
Macarthur, William 3, 5, 11, 93, 111, 113, 164, 166
Macdonald, John Murray 175–77, *176*, 178, 181, 196, 199 *see also* Ben Ean
McGuigan, Brian 257–59, *259*, 262
McGuigan, Fay 258–59, *259*
McGuigan, Perc 220, *221*, 222
McKenzie, William 125, 167, 170
Macquarie, Lachlan 32–33
McWilliam's 209
    Pyrmont laboratory 246–48
    wine sales 214, *218*, 218–19, 232
    wines from 249–50
    see also Mount Pleasant
Madeira, Portugal *see* Portugal
Madeira (wine) 7, 39, 109, 110, 155, 172, 180–81, 188

Maitland Mercury 102, 103, 105, 109 *see also* John O'Kelly
Maitland, NSW 14, 35, 105–106, 132–33, 142 *see also* Dunmore; Eelah; Poole Farm; Telarah
Malbec 133, 166, 174, 180
Manual of Plain Directions for Planting and Cultivating Vineyards 8–9, 47–48, 60, 64, 69, 83–84
Manuel, HL 202, 205
Margan, Frank 251
Marsanne 96, 112, 242
Matheson, Digby 244
Maxwell, Allan 232
medicine, and wine 126–27
Merton, Denman, NSW 56–57, 98–99
Morpeth Wharf *140–41*
Mount Pleasant, Pokolbin, NSW 20–21, 208–209, 223, 225, *226–27*, 246–47, 249–50 *see also* McWilliam's; Maurice O'Shea
movies 246–47
Muddle, John and Mary 258, 261
Munro, Alexander 123–25 *see also* Bebeah
Munro, Sophia 124
Muscat 17–18, 60, 155, 170
Muscatel 60, 63, 74
Muswellbrook, NSW 63–64

N

naming conventions 7, 19, 39, 164–65, 243
networks
    social 49, 57, 71, 79, 81, 100–101, 150, 178–81, 213–14, 262–63
    vine exchange 50–51, 60, 63–64, 112, 130–31
    see also Hunter River Vineyard Association; Pokolbin and District Vinegrowers Association
Newcastle 32–33, 36, 151, 189 *see also* Ash Island
New South Wales Agricultural Society 44, 93, 111
New South Wales, colony of 4–5, 32–33, 43

New South Wales Legislative Council 69, 104–105, 108
newspapers 103, 197, 249, 251 *see also* *Maitland Mercury*
Nicholson, Arthur 169–70, 172–74
North British Company (NBC) 119–20
Northumberland Hotel, Maitland 106, 132–33
Norwood, Allandale, NSW 174–75
Nowlan, JR 154
Nowlan, Timothy 49

O

Oakdale, Pokolbin, NSW
    see Wilkinson family
O'Connor, Tamsin 32
Ogilvie, William 56–57, 139
O'Kelly, John 109, 131–32
O'Shea, Léontine 208
O'Shea, Maurice 208–209, *208*, 219 *see also* Mount Pleasant
Oporto 50, 74, 109
Orindinna, Gresford, NSW 114–15
Osterley, Raymond Terrace, NSW 92–93

P

Page, Geoff 245–46
Park, Alexander 60, 68–69, 108
Parnell, Montague 92, 105, 168
Parry, Edward 82–87
Patch, Thomas 92, 114
Paterson, Jenny 143, 156
Paterson, NSW 33, *34–35*, 133 *see also* Caergwrle; Gostwyck; Tocal; Trevallyn
Patrick's Plains *see* Singleton
payment in kind 103, 106, 107
PB Burgoyne & Co 169, 174
Penfold's 169, 182, 197–200, 220–22, 232, 254, 257–58
Petit Verdot 97, 166
Phillip, Arthur 4–5
Phillips, James 49
phylloxera 163–64, 168, 201–202, 207
Pickering, Denman, NSW 57
Pike, John 47, 49, 57, 63

Pilcher, Henry Incledon 48, 64–65
Pineau 109, 110, 155, 180
Pinot 17, 18, 109, 146, 170, 172, 174, 215, 216
Pokolbin, NSW 130–31, 133, *179*, 184, 210 *see also* Ashmans; Bellevue; Ben Ean; Brokenwood; Glen Elgin; Ivanhoe; Lake's Folly; Mount Pleasant; Somerset Views
Pokolbin and District Vinegrowers Association (PDVGA) *193*, 193, 196–97, 225
Poole Farm, Maitland, NSW 92, 168
Porphyry Point, Seaham, NSW 87, 157, 167, 172–73, 199–200
port 6, 39, 155, 166, 188, 199, 213, 221
Port Stephens, NSW 16 *see also* Tahlee; Tanilba
Portugal 2, 5, 6, 39, 43, 68–69, 75, 88, 126, 163
  grape varieties 50, 63, 74
powdery mildew 120–21, 163, 202

R

Raymond Terrace, NSW 145 *see also* Irrawang; Kinross; Osterley
Redding, Cyrus 100
Reynolds, Charles 92
Riggs, Iain 263–65, *264*
Rolls, Eric 231
Rosemount, Denman, NSW 154–55, 160, 166, 169
Roseworthy College, South Australia 187, 257, 264
Rothbury, NSW 184 *see also* Caerphilly; Daisy Hill; The Wilderness
Rothbury syndicate 252
Rousanne 96, 112
Ryan, Phil *247*, 247–50

S

saccharometers 117 *see also* wine making, equipment
St. Anne's Church, Gresford 200
St Helier's, Lochinvar, NSW 57–59
Sale of Colonial Wine (Cider and Perry) Act, NSW 116, 151

sales 101, 103, 113–14, 125, 167, 169–70, 172–74, 178–79, 232 *see also* wine shops
Saunders, John Morrison 107
Schmid, George 55, 150, 157
Scone, NSW 57
Scott, Alexander 'Walker' 51, 94, 139
Scott family 51, 94, 147–49
Scott, Helenus 51, 94, 96–97, 136, 137 *see also* Glendon
Scottish Australian Investment Company (SAIC) 119–20
Scott, Saranna 136, 137, 147–49
Seaham, NSW 65 *see also* Porphyry Point
Semillon 17–18, 60, 123, 146, 188, 215, 239, 240, 250, 265 *see also* Shepherd's Riesling
settler capitalists 8, 36, 40–41
Shepherd's Riesling 60, 109, 110, 155, 169, 172, 173, 174
Shepherd, Thomas 60, 70
sherry 6, 74, 155, 165, 221, 250
Shiraz 17–18, 96, 123, 146, 155, 170, 172, 180–81, 213, 219
Shortland, John 32
Simon, André 240–41
Simpson, James 115, 158
Singleton, Benjamin 124
Singleton, NSW 14–15, 31, 124 *see also* Bebeah; Glendon; Kirkton; Luskintyre; Windermere
Sissingh, Gerry 219
Smith, Adam 87–88
soil
  composition 18, 27–28, 41, 50, 79, 95, 170, 172, 173, 174
  for grapes 18–19, 42–43, 211, 215, 216, 242
  preparation 46–47, 79, 95, 144, 211
soldier settlers 206–207
Somerset Views, Pokolbin, NSW 255–57
Sommers, Brian 19
South Australia 169, 187, 191–92
Spain 2, 5, 6, 63, 96
Squeeze A Flower (film) 246–47

Stockhausen, Karl 230, 238–39, 256
sugar
  added to wine 84, 85, 196
  levels in grapes 11, 117, 133–34, 176–77, 213
Sweet Water 17, 62, 63, 98
sweet wines 39, 188, 196, 205, 206, 215–16
Sydney Botanic Gardens *see* Botanic Gardens

T

Tahlee, Port Stephens, NSW 15, 82, *82*–83, 85, 85 *see also* Australian Agricultural Company
Tanilba, Port Stephens, NSW 81
tariffs 86–87, 108, 126, 197 *see also* free trade
Taylor, John 139, 144
Telarah, Maitland, NSW 64–65
temperance 126, 203–206, 219, 232 *see also* drinking culture
Terrier, Philobert 120
terroir 19
Thomas Hardy & Sons 169, 198
Tinta 18, 50, 60, 109
Tocal, Paterson, NSW 46–47, 49–50, 76, 92
Tokay 18, 39
Tomago estate, Tomago, NSW 89–90, 112, 113 *see also* Maria Windemeyer
tourism 244–46, 251–52, 263
Townshend, George 48, 59–61, 91, 108
Treatise on the Culture of the Vine and the Art of Making Wine 6, 51, 64, 131
trenching *see* soil, preparation
Trevallyn, Paterson, NSW 59–61, *61*
Tulloch, Hector 219, *223*, 224, 225, 229 *see also* Glen Elgin
Tulloch, John Younie 177–78, 181, 184, 196, 205, 216 *see also* Glen Elgin
Twopenny, Richard 165–66
Tyrrell, Edward 122–23
Tyrrell, Edward George Young (Dan) 123, 196
Tyrrell, Murray 265–66, *267*

V

Verdelho 69 , 109, 169, 170, 172, 173, 174
Verona, Denman, NSW 260–61
vertical integration 115, 233–34
vinedressers 137–39, 142–43, 155–56
vineyards
    costs 38, 42, 107, 167, 181
    early plantings 5, 9, 11, 15, 48–49, 79
    numerical data 104, 183, 282–85
    planning and layout 48, 55, 94–95, 119, 133, 154, 211–12
    work in 46–47, 76, 144, 148, 211, 238, 255
    see also grape vines; vinedressers
viticulture manuals 8–9, 64 65, 75, 76, 239 *see also* James Busby
Vitis vinifera 26, 37, 163–64 *see also* grape vines

W

Walsh, Brian (historian) 50, 54
Walsh, Brian (wine grower) 250, 255
Wantage 60, 98
Warren, Alexander 48, 49, 65
Webber, James 1, 3, 46–7, 49–51, 60, 75–77
Weismantel, Richard and Catherine 152–53, 160
Wentworth, William Charles 88–89, 105, 108, 139
Wenz, Johann 144
Wilderness, The, Rothbury, NSW *121*, 121–22, 168, 170
Wilkinson, Audrey 178, 179–80, 196, 202, 211–13, 225
Wilkinson family 128–31, *129*, 169, 182
Wilkinson, Frederick 125–26, 128
Wilson, Caleb 76–77
Windermere, Singleton, NSW 88–89, 112
Windeyer, Archibald 90, 107–108, 110, 139
Windeyer, Maria 89–90, 113–14, 139
Windeyer, Richard 89, 104, 139
Windeyer, William 89–90, 113
Wine Adulteration Prevention Act 196
wine drinking *see* drinking culture
wine globalisation 19, 158–59, 185, 264, 265–66
wine industry 44, 187, 197–200, 214–15, 252, 254
    vertical integration 115, 232, 234
wine making
    process 70, 167, 187, *208*, *209*, 225, 228, 249
    descriptions of 83–87, 96–98, 222–23, 238–39
    equipment 83, 96, 97, 117, 153–54, 183, 185, *186*, 214, 225, *254*, 265
Wine Overseas Marketing Act 216
Wine Overseas Marketing Board 216
wine regions 18–19, 22, 40, 117, 119, 165, 273n38
wine shops 115–16, 151–52, 153, 204–205, 217–19
wine styles 11–12, 39, 187–88, 205–206, 215–17, 234
wine tastings 108–10, 245
women
    wine drinkers 234, 244
    wine growers and makers 47, 90, 113–14, 258
working class immigration 137–38, 143–44
World War I 203, 204, 206
World War II 220–221
Wright, Samuel 47, 63–64
Wybong, NSW 257–58
Wyndham, George 8–9, 47, 48, 69–71, 91, 94–95, 105
    vine exchanges 50–51, 57, 60, 63–64 *see also* Dalwood
Wyndham, John 115–17, *118*, 125–26, *127*, 132, 161, 166–67 *see also* Dalwood

Y

Yore, Keith 260–61

A NewSouth book

Published by
NewSouth Publishing
University of New South Wales Press Ltd
University of New South Wales
Sydney NSW 2052
AUSTRALIA
newsouthpublishing.com

© Julie McIntyre and John Germov 2018
First published 2018

10 9 8 7 6 5 4 3 2 1

This book is copyright. Apart from any fair dealing for the purpose of private study, research, criticism or review, as permitted under the Copyright Act, no part of this book may be reproduced by any process without written permission. Inquiries should be addressed to the publisher.

 A catalogue record for this book is available from the National Library of Australia

ISBN 9781742235769 (hardback)
     9781742248738 (ePDF)

Cover design and internal design concept Peter Long
Design Susanne Geppert
Front cover image Hector Tulloch, Glen Elgin, c1949. Photograph by Max Dupain. Tulloch Family Collection, courtesy of J.Y. Tulloch.
Back cover image Pokolbin, c1949. Photograph by Max Dupain. Tulloch Family Collection, courtesy of J.Y. Tulloch.
Printer 1010 Printing International Limited, China

All reasonable efforts were taken to obtain permission to use copyright material reproduced in this book, but in some cases copyright could not be traced. The author welcomes information in this regard.

This book is printed on paper using fibre supplied from plantation or sustainably managed forests.